A View from the Heart

Bayou Country Ecology

A View from the Heart

6-26-93

574.526
K

B&T 19.95

A View
from the Heart :
Bayou Country Ecology

June C. Kennedy

Photographs by
Dennis Siporski

Illustrated by June C. Kennedy

Blue
Heron
Press
Thibodaux, Louisiana

A View from the Heart

No part of this book may be reproduced in
any way without written permission from the publisher.

A View From the Heart
Bayou Country Ecology
© 1991 June C. Kennedy
All Rights Reserved

All Photographs
© 1991 Dennis Sipiorski
All Rights Reserved

Cover Design
© 1991 Blue Heron Press
P.O. Drawer
Thibodaux, 70302
All Rights Reserved

Manufactured in the United States of America
Library of Congress Catalog Card No. 91-90461
ISBN 0-9621724-4-8

This manuscript was proofed for scientific accuracy by
Dr. William Barr, PhD., Department Head,
Curriculum and Instruction, Nicholls State University
Former Science Education Supervisor, State of Louisiana
Owner, Science and Environmental Education Consultants

Dedication

For my grandchildren,
Lynn, Ann and Mark

Acknowledgements

"...as I went down the beach I could feel the drawing in men's minds, like, the lowering and shifting realm of color in which the star thrower labored.

...we have kept, some of us, the memory of the perfect circle of compassion from life to death and back again to life, the completion of the rainbow of existence."

*– Loren Eiseley in **The Star Thrower***

There have been many wonderfully helpful people who have given of themselves to help me on my way, many star throwers I would like to thank. From my early days Mr. Arthur Black, Doris Kling, John Pendergraft. Later Tina Eucks, Rick Wilson, Thelma Kiger, and Daisy Guidry. Most recently Damon Veach and Dr. Glen Conrad.

Last, but not least, from the Blue Heron Press, Susan Clement, and Robert Barrilleaux, and especially Carolyn Portier Gorman - my editor, who because she believed, valiantly fought her way through my immense and tangled jungle of words to create a worthy book.

My cup runneth over with stars and gratitude.

VI

Introduction

We must save and sustain and restore our wetlands. But we cannot save what we do not love. And we cannot love what we do not understand.

Growing up in South Louisiana during the Eisenhower era, childhood was dominated by two influences. One was oil — every family had at least one individual who derived their livelihood from "black gold." The other was the mix of land and water that produced the food and recreation that contributed to the uniqueness of the area. Fishing, crabbing, catching crawfish, oysters or shrimp were daily fare. I actually had crawfish in my back yard and they were frequent guests under our carport. Conversation seemed always centered on the oil industry or fishing, or both. Oil and its images were everywhere. Oil field company names on company cars, on television news and advertising, on caps, cups and pictures hung in living room walls; oil was part of life's fabric. We never stopped to consider whether oil jobs and oil itself were limitless.

Bayous and marshlands were everywhere and we played constantly in them. Snakes, alligators, egrets, nutria, herons were just a part-and-parcel of the neighborhood scenery. Boats were as common as cars; many families that I knew owned more boats. Everyone discussed their latest favorite spots for fishing, crabbing, hunting. We believed that the marsh/bayou mix was so limitless that we did not make much fuss when some carelessly used them for "open dumps" and others drained or filled them in saying "they were almost useless wet — except as home for crawfish and water snakes." We believed then that the marsh and her seafood production were endless. The marsh extended indefinitely in so many directions that no one stopped much to think about its limits. My family was always getting or giving away surplus "catch" from some morning excursion. The easy neighborliness was lubricated by the sharing of the overabundant shrimp, crawfish, several kinds of fish, game, etc. We grew up with a continuous supply of "delicacies" that we did not know were un-obtainable most everywhere else.

Oil is no longer king in our home town although his influence still is formidable. We are currently mining the last half of what is easily extractable of that resource. More importantly, we are noticing that the green gold of our marshes and bayous is washing away just as the oil is diminishing. It took a while to realize that this largest concentration of our nations wetlands is melting away — it took a while because it is so vast. At its peak, Louisiana's coastal wetlands measured between 4 and 5 million acres.

In this amazingly complex area, some new wetlands is always being built, some is being moved around (the remnants of our barrier islands are constantly migrating), and some is always being destroyed. The net result of all of these interacting forces, however, is that an acre is being lost every fifteen minutes; all day every day — nature never rests.

Some of the forces sand-papering-away this natural cornucopia of production are labeled subsidence, compaction, erosion, salt water intrusion. But the variety and complexity of the forces eating away these wetlands makes solutions very complex and very expensive. That makes the delta also a fertile area for research and home to hundreds of human experiments trying to control, cooperate with or co-create with these natural forces. Many of our best scientists and engineers are literally waist-deep in trying to figure out how to sustain this system. What we don't know is an awful lot and we know that we don't know nearly enough. But two things stand out amidst the complexities.

First, we know that this wetland system will not sustain us in the future if we do not begin a huge effort to sustain it now. The system that is 40% of our nation's coastal wetlands and produces

A View from the Heart

comparable amounts of her fish, fur and water fowl. We know that this ecosystem will collapse sometime in the next several decades if the "tide is not turned on coastal wetlands losses." Fortunately, thousands of individuals are marshalled in the battle to save the system and the state and our nation are investing about $60 million annually to back them. But they are yet not enough, nor is the money sufficient. They are still losing badly in most areas where it is important and more homes, businesses and jobs are threatened each day. Measure the economic threat alone in the billions and billions. Anyone who cares about the future, or the future of our state, should pay close attention to how this turns out. If too many of us take for granted our shrimp, crawfish and fin-fish, they will be gone.

The second thing that we know is that without the intervention of man, the architect of these wetlands would still be adding to her net total of acres instead of quietly watching them disappear. The Mississippi River was the major architect and she dumped her mud collected from draining half of our continent onto this expanding delta for the last several thousands of years. The Mississippi River is now belted in and we dump her mud off of the continental shelf. Unless we can recycle significant amounts of that suspended mud into these wetlands, they will continue to sink and wash away. Our marshes and wetlands must be fertilized to survive. Otherwise, they squeeze their excess water out (in settling) and sink under their own weight. While recycling mud that we are now letting drift away out into gulf waters is not an overwhelming engineering problem, it is an expensive one. But worth it when we consider the billions of dollars of value produced by the system each year.

There is so much that our scientists and engineers do not understand, even concerning what seem to be the simplest methods of sustaining this wetlands system. The tougher job of restoring and rebuilding this system, a job that the Netherlands have done much better than we have, is still out of our grasp — although closer to our reach. When we try to decide what to do, we notice giant gaps in our knowledge and information. Remember, we cannot save what we do not love and we cannot love what we do not understand.

June Kennedy's book, *A View From the Heart*, combines the magic and mystery experienced in the wetlands, which if nurtured and re-experienced grows to love, along with easily digestible explanations of the general truths and physical phenomena of our wetlands. She gives us poetry and science. That makes it not only interesting to read but also something to share by reading parts out loud to our children. It is a book to pass around the household or give to the neighbors.

While it will take love and understanding to keep our wetlands, *A View from the Heart* gives us plenty of both. Read the first three pages and taste the strong affection and attraction for this magical place that Kennedy evokes. Meet the native American Chitmacha Indians, soar with the Southern Bald Eagle or reflect on the two red dots that reveal the "Lords of the Marsh" — the ancient alligator. And Kennedy ends this first part, the part where she says metamorphosis began with this: "Surely we will come to see the beauty, wonder and perfection of such wilderness as a precious link to the past. It can also be an equally priceless link to the future if we care for it as one would an irreplaceable treasure." Her vision is a sensibility and a rapport devoutly to be wished.

June Kennedy's writing captures the spirit of the marsh, our liquid land as she calls it, and the poetic quality of her writing can sustain the reader through absorbing the scientific information she artfully blends in. Information is the basis of understanding and Kennedy makes information bite-size and tasty. We need more books about our wetlands written this way!

To lose this wetlands system is unthinkable. Our grandchildren must have it, if only for the production it could yield. We know that they will not have easily extractable oil. To sustain and then restore this system we must love it better than we have and understand it much better than we do. June Kennedy gives us avenues to both increased appreciation and deeper understanding. I am grateful that she produced it and I had a chance to enjoy it. I hope you will share it with your neighbors, your children and grandchildren.

- Michael Mielke
Executive Director
Coalition to Restore Coastal Louisiana

Contents

A View from the Heart

1

The River to Bayou Country

There wasn't the slightest hint, our last night on the river, that the next day we would begin to discover a whole new world that would change our lives forever - Bayou Country. We would not be aware at first that we were becoming part of it, even as it was becoming part of us. In time, its people would become our people; its beauty, a natural high that feeds the soul. We were on the Atchafalaya River when the metamorphosis began...

Atchafalaya River
Mile 304.2
Journal Entry
November 1, 1977

This has been a delightful evening, one of the most pleasant we have spent on the river. The setting is perfect. We are moored beside the wall of Old River Lock in a quiet spot out of the channel. The long lock wall provides an excellent place on which to bring out our little hibachi and grill our supper. On the down river side, the setting sun puts on a particularly spectacular show, which the water reflects. The fish jump and hop in the colorful water, providing music complete with lighting effects. The evening is balmy and the cement on which we sit still holds some of the sun's warmth. On the up river side, tows and barges move very slowly through the narrow approach to the lock. They are so close we could easily touch them. As they slip along with a quiet swish of the water, we are fascinated by the fact that they are moving through a wall-enclosed space with only inches to spare on either side. Rivermen call this "shoe-horning in." We nod and wave to the pilots and chat with the young men always present on the lead barges.

A View from the Heart

Time slips by as quietly and pleasantly as do the tows and barges. We retire and all night long the gentle swishing reminds us the river traffic never sleeps ... but we do presently and pleasantly.

Believe me, it was not love at first sight - it was more like shock. We had become accustomed, on our journey down the Ohio and Mississippi rivers, not only to wide rivers, but to having them well-marked with lights, buoys, channel and mile markers - the whole nine yards to help us. We were ill-prepared for the narrow, completely unmarked Atchafalaya River. The only signs we saw were warnings: "No Trespassing," "No Hunting," "No Dredging-Buried Cable." Our NOAA (National Oceanic Atmospheric Administration) map, also showing no markers, was all we had to guide us. Only by using tremendous concentration to figure exact time and speed, then actually making a pencil line along the map using these figures, plus our depth meter, were we able to navigate safely. There were tough decisions to be made when reaching a point where there were various forks to chose from. All we could do was hope the fork we could see was the same one we were looking at on the map. It was like being in a living maze. Only much later would we learn that people were advised not to travel the river without an experienced guide. No one had even remotely warned us of what we were getting ourselves into.

By nightfall, without the foggiest idea of where we were, we moored to a tree, ate a cold supper, and just sat - it hadn't been a good day. We have since discovered the Atchafalaya is home to 65 species of reptiles and amphibians, 46 species of mammals and 90 species of fish, shrimp and crawfish. I think every single one of them came to visit us that night, for the air and water were alive with sound and motion. The two red dots in the water that marked the presence of several "Lords of the Marsh" could be seen in the pale light of a near new moon. It was exciting, but a bit scary too. The coming of the dawn gave us a first whispered essence of the amazing beauty and profound serenity of this river basin.

By afternoon of the second day we reached Grand Lake. The only other humans seen on the entire trip had been two or three solitary fishermen in little bateaus.

Once on the lake, whatever gods had given us protection in passage decided it was time to awe us with their power. Out of nowhere a storm whipped up with howling winds and driving rain, the like of which we had never seen before - or since. Just as quickly as it started, it was over; the sun beamed brightly, the water was again tranquil. We had the weirdest feeling that somehow we had passed a test and were now being smiled upon and accepted by this living river.

It is an immense place, this massive Louisiana river basin, over 1,000,000-plus acres, the largest of river swamps, and one of the last of its kind, in this country today. Even the name is big - Atchafalaya (a-cha-fa-ly-a). The word glides over the tongue as easily as the Indians who named it glided along their "River Long." Truly it is a long river. Stretching roughly from Simsport in the north to a fringe of watery lace on the Gulf of Mexico, the Atchafalaya River is in reality a huge river basin, 18 miles wide and 135 miles long. It encompasses a wide variety of flora and fauna from verdant coastal ecosystems in the south, to ancient cypress swamp in its heart, to hardwood forests in the northern reaches.

At its southernmost points, Nature, Time and the River are creating a new delta, while across the state, to the east, the same three, with extra help from Man, are tearing down the delta. It is paradoxical, there must be change so that things can remain unchanged. If change stopped, so would life.

Bayou Country Ecology

So this living wilderness changes, and yet in many ways it is much the same primeval wilderness the Chitimacha Indians knew when it was their home. Here the spirit is still free to soar with the Southern Bald Eagle. The eye can still catch, but not hold, the lovely white-tailed doe as she moves hesitantly to the water's edge. The ear can still hear the *swoosh* of an alligator sliding off a log into brown water, or the splash of black bass surprising unwary insects. Over all is "heard" the ambient sound of silence.

Here the night sounds still hush in anticipation of a new day. Here the sun never bursts into view, as it does along the coastal marsh. Instead, light comes slowly, quietly flooding each droplet of morning mist and swamp water, creating an ethereal light that seems to come from everywhere and from nowhere, holding you spellbound till the new day finally wafts through moss-draped trees and dances lightly upon the water. Life moves to a quickened beat. This is the swamp primeval, where creatures of the basin awake to go about the age-old business of survival in the near-silent ancientness of this great cathedral. Perhaps that is why we find ourselves whispering, or is it the feeling that a single spoken word could shatter this world of fragile crystal-like beauty?

One has to spend much time alone in such places to become truly attuned to their particular rhythm, the long rhythm of the days, the nights, the years, the centuries. Flow and movement are constant, yet time seems to be held at bay. You can still hear silent echoes of the long past. You feel that if you would only reach out, you could touch the very Beginning. Man is yet an intruder here in one of the last great oases of true wilderness.

Most of the big cypress are gone now. Those that are here are second growth, mere shadows of the giants that once reigned here. Now and then one comes on a somber old tree, standing like a ghost sentinel from some forgotten time. Ghost, no. Survivor, yes. Passed over by swamp loggers of long ago because it simply did not measure up, this is one of the last. A third generation will probably never be seen here because slowly, almost imperceptibly at times, the same annual flooding that is the life blood of the swamp is filling in the basin's shallow waterways with sediment. Frivolous, wand-like willows are taking over. The willow is an opportunist tree, giggling at every passing breeze, yet needing only a drop of water and a grain of sand to sprout.

Man has posted his signs everywhere here, and for the most part, his fellow men, read them. Nature, too, has posted her signs: ***Do not disturb balance - delicate. One-of-a-kind. These laws can be bent, but not broken.*** Either we cannot read them or we do not care to heed them. Some still do not realize we are locked into the programming of creation. Either we learn her laws and adapt, or like all other creatures that have failed to do so, we too, will vanish, only to join the vast army of species lost forever in the stream of time that feeds the eternal swamp.

Surely we will come to see the beauty, wonder and perfection of such wilderness as a precious link to the past. It can also be an equally priceless link to the future if we care for it as one would an irreplaceable treasure.

2

Bayou Country

*"It is a place that seems unable to make up its mind whether
it will be earth or water, so it compromises."*
- Harnett T. Kane

Bayou Country is true wilderness, most of it virtually untouched by either human roads or inroads. In certain places, it comes close to seeming primeval in nature, as close as it is possible to find this in the world today. In these sections it looks just as it must have looked thousands of years ago. Yet Bayou Country is millions of years younger than even other parts of Louisiana, which isn't itself very old - in a geological sense.

The word *bayou* probably comes from a Choctaw Indian word meaning - small stream. To south Louisiana residents, bayou may mean almost any body of water in the area, except the Gulf itself. It is even used to describe or name dry land places, like towns. Unlike streams, bayous rarely dry up because they do not depend on springs as streams do. Instead, they are a product of an oversupply of water spilling over a submissive soil, ever hunting and finding alternate ways to the sea. Along the way bayou networks create a fantastic sampler of Nature's lace-making skills.

Bayous serve as a drainage system for an area that cannot ever have enough drainage. It is, in fact, a balance wheel maintaining the equilibrium between rivers, lakes, marshes, bays, swamps, and even other bayous. For six hundred miles the broad Mississippi meanders along the border of Louisiana. Four major rivers cut through this swampy lowland, the Sabine, Pearl, Calcasieu, and Atchafalaya. Half of the state's some twenty thousand square miles is below the high water level of these rivers. There are unnumbered hundreds of bayous that move these waters along. They come in all sizes but one wonderful word describes them all - *bayou*.

Like the people who call them home, bayous make their own rules and take their own sweet time when moving. Not set in their ways, bayous may flow in two directions at once. One, flowing into a lake, may, at night, respond to a tug from the moon and flow in the opposite direction, seeming to enjoy the leisure of doing exactly as it pleases. Again like people, bayous come in no set size. They can be deep and powerful as rivers and as wide, or so small, shallow and narrow that a pirogue - a pea shell of wood that can float on dew - can barely slide through. Bayous build their own banks, create their own levees, push outward their own tendrils of land. Despite all these things, bayous are seldom much more than a foot above sea level. Bayou Country itself is no more than a maze of interconnecting bayous, shallow lakes, canals, inlets, cuts, cutoffs and their adjacent levees. It is a liquid land, complex and maze-like where what little soil there is, is dark and rich.

Luxuriant with the silting of soils drawn from half a continent, this is an area warm and fecund as a womb, where rich and varied life, in water, on wing, on paw, and on stem, is as overwhelming in its kind and numbers as are the bayous themselves. Here in a flat world of lonely silence, are many creatures dwelling together that may have lived side by side for millions of years. Some microscopic species may have shared the land with the dinosaurs. Birds that may be using the same flyways they have followed since time immemorial, come yearly to fly side by side with native birds of this warm moist semi-tropical wilderness where death feeds life, and life produces more life in wild abandon.

The echo of the ancient past is still here if we look carefully. A harmless floating stick becomes a water moccasin. A log, drifting along, may well be an alligator. A snake swimming quietly along, suddenly shrieks and takes wing - a "snake bird" or anhinga. That harmless boulder is not a boulder at all but a large alligator turtle. Suddenly you feel as if you could reach out and touch the past. It comes as no surprise that for so long such places were considered evil. True, such beauty and startling contrast can mean trouble for the unwary or careless, but it can mean adventure and wonder to the careful and those willing to learn. Above and beyond all else, bayou country is a fascinating showcase of bio-diversity, the tens of millions species and the billions of distinct populations of plants, animals and microorganisms that share the Earth with us.

Bio-diversity offers a showcase of Nature's creations. Wilderness, untamed land or sea, does by its own nature, support a greater abundance of life forms than developed areas. In Bayou Country both variety and diversity are present. Diversity is Nature's grand tactic for survival in that it guards against complete annihilation from a single source. Ecosystem is the name given to the community of many and varied creatures that interact within a given environment. Ecosystems depend on bio-diversity to maintain their equilibrium. The greater the diversity of an area the less chance for it to be destroyed. Abrupt changes in an ecosystem can only be adjusted to when it is complex enough to allow alternate relationships to develop. For example, if one source of food for a creature disappears from an area, another must be available, otherwise the creature in question cannot survive. So it is easy to see that to lose species from an ecosystem is to weaken it. The study of ecology is basically the study of ecosystems.

Let us take an imaginary trip to either a favorite bayou, or for those who have never experienced that joy, a typical bayou in this region where trees grow tall and Spanish moss hangs in feathery grace. Dark placid water reflects with mirror clarity trees that seemingly go endlessly up and endlessly down - the first of nature's illusions. Just as this image suggests, what we are searching for lies beyond physical seeing perhaps, but not beyond the ability of our mind's eye to understand. This search can bring us to a newer and certainly more lasting sense of the order and beauty that surround us.

We have come to hunt sunbeams. Morning is always a lovely time on the bayou. Gentle light filters through the tall trees in beams that touch the water. Sitting very still, you might see how some plants turn their leaves toward the light and many flower petals will open in it. If you take a piece of glass along, you can use it to see the spectrum of color it creates. Consider how this messenger of the sun brings warmth and light across 93 million miles of cold black outer-space. Then perhaps not only will you see the beauty of the light, but you will get a "mind's-eye view" of the world of the radiant energy of color.

The same light that allows you to see the morning may call into being a marsh flower whose fragile beauty will last but a single day. Similarly, a myriad of winged creatures of the marsh, who before the sun has set, will sink, wings folded, into death. Each has enjoyed its allotted share of life. The light and warmth of the sun may beckon a small seed from the earth. One day it will tower in the sky, a mighty live oak, braving wind and storm for perhaps hundreds of years before, twisted and gnarled, it too will sink into decay. When each living thing has spent or lost its share of life energy, it dies. Whether in minutes or after hundreds of years in the organic world, life forms return to the world of inorganic

4

Time Remembered

"If there is magic on this planet, it is water. . .
it's substance reaches everywhere.
It touches the past and prepares the future."
- Loren Eiseley in Immense Journey

In the years when the river was free to flood, its freshwater and sediment constantly rejuvenated the coast, even building up eastern Louisiana beaches. But as New Orleans and surrounding areas grew and developed, embankments were built to protect them. Over the years, billions of dollars and untold energy have gone into an ongoing effort to control and contain the Mississippi River, to prevent it from abandoning its present delta course. Levees and spillways now are relatively successful in protecting all inhabited areas from floodwaters from the river.

The constraint of the river was necessary to human development and business. But at what cost? The effect on the health of coastal ecology has been dramatically severe. The river created this delta, and when its life-giving sediments were no longer allowed to spread over the marshes and shallow coastal waters, even at flood time, it was only a matter of time before fresh-water plants began to starve, not only from a lack of water but from a lack of nutrients as well.

Fresh-water marsh typically supports vegetation that cannot tolerate long exposure to salt water. They quickly become "stressed" and trees, plants and grass soon die. With roots no longer there to hold the soil together, it soon washes away before more salt-tolerant vegetation has had time to gain a foothold. Although brackish water environments suffer somewhat less than do the fresh-water environments, any type of marsh loss is disastrous.

For man, development could not stop, for business would not stop. Oil had been discovered in Louisiana in 1920; "Black Gold" was the name of the game. Oil was discovered in Leeville near Grand Isle in 1930. The 1940's saw the first offshore oil wells along Louisiana's coast.

The effect on the economy was dramatic. Grand Isle, a sleepy barrier island resort and fishing village, became a busy, thriving place: "Oil City" on the one end and new beach resorts on the other. More building on the beach side made the isle increasingly vulnerable to hurricanes. Why? Because all healthy sand beaches maintain a delicate balance of sand supply, beach shape (or profile) and wave energy. Any construction on or near the shoreline changes this balance and reduces the natural flexibility of the beach. The balance is so delicate that even beach cottages on stilts may obstruct normal exchange between beach and shelf during storms. This balance was completely ignored, however, and the development of Louisiana's coast line accelerated.

Channels and canals were built; marsh buggies reigned. Eight thousand miles of pipeline crisscrossed our marshes. "Times were never better". . . and so it went for many years.

21

A View from the Heart

In the late 50's, the Mississippi River Gulf Outlet was begun. By the late 60's, anyone in St. Bernard Parish could have told you what was happening. The river's life-giving water and sediment was being funneled downstream, past the marsh and into the deep waters of the Gulf of Mexico. The outlet was already causing thousands of acres of marshland, present in the 50's, to be lost. Salty, dead bays and lagoons replaced estuaries that had teemed with shrimp and crab. It was not long before everyone in the state knew we were losing more than 40 square miles of coastal wetlands every year. With these acres went a considerable part of the state's fur production area. Considering that at the present time Louisiana is still at the top nationally in the production of both fish and fur, what a bountiful natural haven of plenty this area must have been in the early days. The Army Corps of Engineers, and other scientists, were soon saying that at the current rate of erosion the shore line of the Gulf of Mexico would be above the Intercoastal Waterway in fifty years or less.

Thirty years ago many believed coastal erosion could be stopped by some type of control structure or system. What was done along parts of the Atlantic Coast was to create what is called the "Concrete Coast" - an ugly, sprawling, constantly growing concrete wall at the shore's edge. This "New Jerseyization" of the coast continues to cost millions of taxpayer dollars for only partially satisfactory results. On the Pacific Coast, sand replenishment was tried, without long-lasting results, despite a high cost. Today, however, more and more places are following the lead of Miami Beach, a barrier island on the Atlantic Coast, and replenishing the sand yearly on a parallel with the repair of their roads. It is now considered part of the standard upkeep. This works well for places that do not mind heavy taxes, but not too well for areas with money problems.

It is amazing that so little was known or understood about coastal ecology. Unfortunately this time period in the development of the area is a perfect example of what happens when human productivity and Nature's productivity meet head-on. Nature loses, but in the long run so does man. We need keep in mind that while this time period seems to us extensive, the human time frame is very short-sighted. A year is considered "a long time." But for Nature, whose time frame is very long, a hundred years is barely yesterday.

No really appreciable improvement has yet been made. All agree that something must be done to stop marsh loss and coastal erosion. But who exactly has the necessary wisdom and knowledge to know if what we do would not, in the final outcome, simply compound our errors. That is not only true of our coast but of all coastal areas in this country. Louisiana, and the rest of the nation, is facing a crisis in its marshes of horrendous proportions.

We do know we can not stop change and that we may never be able to. Perhaps for now, till our knowledge and wisdom as a people are greater, the big question should be not how to stop change, but what can we do that will do the least harm and buy us the most time. As writer Wendel Berry so sagely points out: "We cannot know what we are doing until we know what Nature would be doing if we were doing nothing." It would be practically impossible to over-stress the wisdom of this statement.

What does seem like a step in the right direction is that vast efforts are being made in the study of coastal ecology - efforts to really learn to understand Nature's ways. If we can learn to understand Nature and learn to work with, instead of against, her plan, perhaps we can find a way to at least stop the accelerated rate at which the Gulf of Mexico is winning and Louisiana is losing the battle for coastline and marsh.

This battle, as stated previously, is not taking place only in Louisiana. In some parts of our country, coastal residents are attempting to find new patterns for living on barrier islands. Over-development, and the natural fragility of some of the islands are given as the main cause of barrier island erosion and other coastal problems. On those islands considered extremely fragile, any type of construction is forbidden at all, and, in a few instances people are forbidden as well.

The Federal Government has taken some steps on its own. As of October, 1983, property owners are no longer able to get Federal Flood Insurance for dwellings on most barrier islands. Some few, such as Miami and Grand Isle, were exempted. The ban did not prohibit insurance, but simply made it more expensive to obtain, thus hopefully discouraging extensive expansion. Perhaps it has served to make the public somewhat more aware of the plight of our country's approximately 295 barrier islands.

In the old and ongoing attempts to save Grand Isle, Louisiana's only inhabited barrier island, work was begun in December, 1983 on a hurricane protection levee and a beach erosion project. This involved building a sand levee which was eventually planted with appropriate vegetation. Access to the water was via raised pedestrian walkways. Early in 1985, things were looking good for Grand Isle. The beach grass was beginning to grow. Then, unprecedented in recent history, Grand Isle was hit by not one but three hurricanes in the same season. During the first two hurricanes, the beach eroded in excess of 100 feet. After Juan, the third hurricane, approximately 20% of the levee was destroyed. Admittedly, the western end of the levee was damaged because a sand borrow site had been dredged too close to shore. The pedestrian crosswalks became "bridges to nowhere." The Grand Isle levee never had time to survive the initial adjustment period required after construction of any such structure to establish a new equilibrium and, eventually, accept a new beach profile. During this period of time, Nature adjusts the man-made construction to suit herself, and then accepts a new beach profile. This requires at least a year, often longer.

Since the beach replenishment system is considered by most authorities the best method of beach protection, as of the summer of 1990 it was being tried again. Hopefully our luck will be better. Engineers are mixing clay with sand this time, since, admittedly again, they know that the sand used last time lacked sufficient amounts of clay. To date, there has only been one mishap this time: the sand dredge sank in rough water, the top part of the dredge came off and ended up nearly on the beach. During the first six months of 1991, unprecedented in Louisiana history, more rain had fallen than ever before. With the rains came high tides and more beach erosion, threatening homes and the only highway that runs the length of the island.

The Army Corps of Engineers has worked at a fantastic rate of speed to be able to save most of the "endangered" camps and the highway. If there are no hurricanes this season, the beach may have the time it needs to establish a new profile. At the moment one might say it is a race between man and the elements. No one can say for sure who is going to win this battle. But there is no question that either way the war will continue....

Before attempting to work out the problems of erosion, we should be aware of some theories about barrier islands. There is a widely-held theory that states that barrier islands migrate landward by an intricate mechanism - that what appears to be erosion is really a natural, inevitable shift of the entire island landward. Those who hold this theory state that misguided efforts to stop erosion and, therefore the island migration, may endanger the island more than help it in the long run. They say that even an experienced observer may not notice the magnitude of such retreats - washing away at front, adding on to the back - because a barrier island tends to assume the same shape as it moves. This manifestation is called "Brunn's Rule."

There is another theory that supports the idea that the world-wide rise in sea levels is causing the barrier islands to be inundated and "sink." These are, admittedly, theories only, but we still need to be aware of them.

There are, however, a few guidelines concerning coastal and barrier islands that are practically foolproof. They have been established through trial and error by knowledgeable people world-wide.

• First, the main guiding principle, whenever possible, must be "hands off." As little interference as possible is necessary to allow coast and marsh to achieve its own equilibrium.

• Another guideline suggests that each type of coastal environment has its own unique problems.

A View from the Heart

Anything that is done, no matter how insignificant, to a barrier island has strong ramifications that effect all parts of the surrounding ecosystems.

We cannot stop change. Even a fifteen foot concrete wall built around a barrier isle cannot keep it safe forever. We must remember that everything moves in a slow, but steady, flow. For conservation measures to be truly successful they must take this into consideration and be truly anticipatory.

One important element to consider when dealing with barrier islands is storms. Regardless of how damaging they may seem, storms help renew life and establish timetables for succession - the natural pattern of change in an area over hundreds of years. For example, storms often provide for the continuation of a species that might have otherwise disappeared. This "reshuffling" by Nature, causes life to renew itself and creates a fresh equilibrium. Many geologists believe that storm conditions are completely responsible for the ongoing redistribution of shelf sediment which is needed to create and nourish beaches. If indeed this is the case, then storms are truly beach benefactors and not the enemy. This is hard for us to accept. Nature has long been considered "the enemy" in these situations. Perhaps now it is time we started treating Nature like the friend she is, and attempt to understand her ways.

Of course to be truly successful, restorations of our coast must include restoration of our marsh as well. They are interdependent.

Few question the part a lack of freshwater plays in marsh loss. There is also no question in the mind of anyone who even remotely understands Nature's rules for salt water tolerance, that salt water intrusion must not only equal but probably exceed fresh-water loss as a cause for marsh loss.

Presently certain companies are attempting to convince the public the loss of marsh is caused more by levee construction than by salt water intrusion via canals and channels built by these same companies. One company recently claimed the levees broke a natural cycle and Nature, who routinely builds and destroys marshland on her own, when faced with these levees, is stuck in a "destructive groove." This idea was expressed in a very slick, free, full-color brochure. Part of the purpose of this book is to provide information that can counter or balance erroneous material like this that are presented to the public, at great expense, as facts.

What more is there to say of coastal and wetlands loss? Coastal wetland loss is double trouble because the wetlands serve as a basis for a food web that sustains fishery resources. The wetlands, of course, provide critical and crucial nursery habitats for future generations of seafood in the Gulf. If coastal wetland deterioration, or sudden loss, can not be effectively stopped, it is only a matter of time before Louisiana's stature as the nation's top seafood producer will end. This equation explains it:

rapid erosion = increased detritus
increased detritus = a <u>temporary</u> surge in seafood production
wetlands losses finally takes place due to rapid erosion = a sharp decline in seafood production

Bayou Country Ecology

As recently as 1985, Mark Chatry, then supervisor of Grand Terre's Marine Lab, wrote an article titled "Hope for Our Coasts." He points out: "In searching for a solution to marsh deterioration and recognizing that salt water is very much to blame, one recognizes that there are only two options, restrict the flow of salt water from the Gulf or direct the river water through the levees into the coastal area." A recently published booklet by the Coastal Restoration Foundation furthers this point: "In order to stop wetland loss and rebuild the marsh, sediment in addition to fresh-water must be reintroduced on a very large scale. Another method for creating marsh involves the pumping of material dredged from navigation channels into the marsh." The booklet states: "Activities being undertaken by Louisiana and local government to preserve and restore wetlands include vegetable planting, fencing and vegetation, revetments (embankments) and offshore segmented breakwaters."

Some marsh restoration progress has been made; some projects have been successful, but all such projects take large amounts of money. Our marshland is vast. This is no easy problem, and here in Louisiana, and elsewhere, saving the country's wetlands must be top priority.

The New Orleans District Army Corps of Engineers recently released figures showing the average loss rate for the entire Louisiana coastal zone has dropped from the roughly 40 square miles per year of the late 50's to early 70's to roughly around 30 square miles annually from 1974 - 1983. This is, of course, good news, but what does it suggest? It suggests things are always in a state of change - Nature is never still. It suggests that as traffic in the oil patch decreased, so did the loss of wetland. This may mean hands off, whenever possible, is often the best policy. Finally, this drop in land loss points out the need to further study Nature's ways. There has been no indication of an overall rebuilding of wetlands, only a decrease in wetlands loss each year. Thirty square miles per year is still too much. While it does hint at Nature's ability to endure, these findings do not change our role in preserving and restoring wetlands.

Are there other things we need keep in mind as we attempt to rebuild and restore the wetland. Yes, we need to remember what John Muir said: "When we try to pick up something by itself, we find it hitched to everything else in the Universe." Francis Thompson, English poet, put the same idea this way: "One can hardly pluck a flower without troubling a star."

There are many examples of the truth of this. A simple and obvious one is the interconnection between creatures in an ecosystem. A less obvious, but interesting example, happened in 1988 when salt water from the Gulf traveled up river to New Orleans. This was caused not by anything here, but by a drought in the Midwest which reduced natural fresh water in the river, allowing a salt wedge to move along the floor of the river from the Gulf to New Orleans. Interconnections like this in Nature are something we need always keep in mind as we plan ahead. So often we do not see these interconnections, except in hindsight.

The time frame for environmental damage is long. Not enough people care about the next hundred years. Too many care only about the next couple of elections, the next five years of land development, the time till the boat is paid off, the time till retirement. There is a Chinese proverb - the exact words are forgotten but not the wisdom: If you are thinking in terms of a year - plant a seed. If you are thinking ten years - plant a tree. If you are thinking a hundred years - teach the people.

In these first four chapters we have explored Louisiana from its inception. Along the way we have experienced the truth of Justice William O. Douglas's statement: "Science can displace or destroy, it can interpret, it can imitate, but science cannot take the place of the wonders of creation." Above all, we have learned how lucky we are. Louisiana still has a rich and beautiful natural heritage - a wondrous gift of Nature, Time, and A River flowing. Let us explore it further.

5

Wetlands

*. . ."thus the proper place to hear again the echo of ancient myths
and remember connections that came before our present dreaming.
Why should we care? Some of us do not, but those of us who do
are persuaded that the connections that tremble at the edge of memory
will be the salvation of tomorrow."*
Ancient Lives - *Wilderness Magazine, Spring, 1988*

What are they, these strange minglings of earth and water that from the beginning of time have fascinated and frightened us. An early Roman writer called them "filthy swamps." No less a personage than Hippocrates himself said: "Such waters I reckon bad for every purpose." So it went for hundreds of years.

By the 1600's, such places remained misunderstood and were considered sites needing reclaiming for better use. Fear lingered. By the 1850's fear was becoming mixed with a touch of reverence by certain writers who hinted that such terrible places somehow held a graver and more profound secret. They hid perhaps, the mystery of creation itself. About this time Thoreau wrote: "I enter the swamp as a sacred place, a *sanctum sanctorum*. Here is the strength, the marrow of nature." Still, we did not fully understand.

Even today, we do not fully understand wetlands, nor can we really define them clearly. For whatever reason, for many years we have bulldozed, clean cut, flooded, filled, poisoned, and paved our wetlands. How then can we be sure that while we wipe clean all traces of the past, we are not also wiping away the pattern for the future.

How do we describe wetlands? Here in Louisiana they include tide-swept, saltwater marshes, freshwater marshes, and brackish water marshes. There are brooding swamps of many kinds, estuaries where river currents meet sea tides, and silent cypress swamps. Wetlands are found worldwide and are known by various names: marshes, sloughs, swamps, estuaries, salt lakes, bogs, pot holes, sinks, fens, oxbow lakes, rivers of grass, and duck factories. These wild and wet places are found in many countries - India, Bangladesh, Russia, the Sahel region of West Central Africa, the Arctic Tundra, and even between fissured craters and fields of lava in Iceland. For better, or perhaps for worse, for want of a better word, we call all such places simply - wetlands.

Is there anything these tremendously diversified places have in common? Yes, many things. While wetlands contain a relatively small part of our world's water resources, that part is extremely crucial. The reasons are many and, like everything else in nature, highly diverse as well as attached.

One thing shared by every wetlands area on earth is constant or periodic soil saturation. The soil must be saturated long enough to assume a characteristic called *hydric*. Such soil lacks atmospheric gases for an extended period and therefore can support vegetation, such as bullrushes, cattails, and other plants with "wet feet." There are many ways soil can become hydric. The water table may be brought to ground level. Water, whether from lakes, ponds, oceans, bays, streams of rivers, can bind all the components necessary to produce wetlands; conversely the opposite is not always true. Not all areas containing hydric soil are wetlands. To be wetlands two conditions must be present, a well-defined water regime and "water-loving" - *hydrophyte* - vegetation. It is these two factors occurring together that usually give an area the designation wetlands.

The most recent **Status Report on Our Nation's Wetlands** by the National Wildlife Federation, informs us that wetlands "value" is not synonymous with "function." Functions include what wetlands can accomplish - how they work. They do not reveal to us value.

Wetland functions include the effect these areas have on:

1. Population - plants and animals that require wetlands at some stage of their lives. This category suggests the value of our estuaries to fish and also endangered species for habitat.

2. Ecosystems - processes such as flood control, storm energy abatement, recharge of aquifers, pollution control, and water quality improvement. Of course, wetlands neither do *all* these things, nor do they do them all at the same time.

3. Worldwide - the cumulative biochemical contributions wetlands make to the global environment. It is believed wetlands play a significant roll in the global cycles of nitrogen, sulfur, methane and carbon dioxide. Some remove excess nitrogen from the air. Others may remove carbon. These concepts seem exciting when one considers the amount of atmospheric pollution worldwide today.

We will go briefly into defining the "value" of wetlands in order to understand what is being lost. The economic value of wetlands on a dollar basis is estimated by a complicated process involving dollar value of harvest plus energy analysis. This type of information is not crucial to this writing. What we need to understand, instead, is the value of vital "free services" rendered to us by the wetlands.

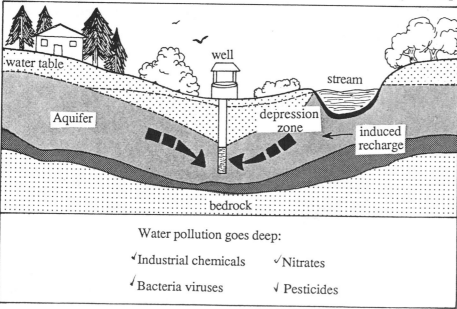

Figure 6

Bayou Country Ecology

As we know, water moves through a cycle and returns to the surface as rain, snow and sleet. Much of it eventually finds its way back to the sea. But a substantial portion seeps deep into the ground. There the water is stored in natural underground reservoirs, which are called aquifers (**Figure 6**). Huge amounts of water have filled these aquifers underlying the continents in quantities far greater than that found in rivers and lakes at any given time. Such water does circulate, but only ever so slowly compared to surface or atmospheric water action. Underground streams and springs return water to the surface and slowly carry it back to the oceans while fresh water constantly seeps down from the surface to replenish the supply.

In some instances, wetlands play an important part in recharging ground water supplies. Whether a wetlands serves as a ground water *recharge* or *discharge* site depends on its position relative to the water table. Wetlands interception on acquifer gradients are likely to be discharge sites, those located above this gradient may recharge the water table. Because water tables can fluctuate with seasonal and climate variation, some wetlands can be recharge areas during dry months and discharge sites during wet months. Either way, we see the important role wetlands play in the cycle of water.

Many wetlands can slow and retain large amounts of water. They are of extraordinary help as temporary storage areas in altering flood flows, thus making floods less damaging. Isolated freshwater wetlands and other non-riparian wetlands (not situated on banks of a river) contribute to flood control because the water they retain almost never reaches watercourses when they are at flood stage.

While science is only beginning to understand the value of wetlands in maintaining water quality, it is understood that nutrient retention and transformation are important functions to consider. The high rate of biological activity, growth, decomposition, nutrient, and energy recycling of the wetlands can result in many wastes such as certain chemical pollutants being absorbed and incorporated into the plants or given off as a gas.

Coastal wetlands absorb and temper the impact of storm surges. Barrier islands, by their very nature, are sparring partners with the sea. But the entire wetlands area can act as a giant storm buffer without sustaining much lasting damage. Coastal development in this area, however, is likely to cause costly storm damage if the normal and critical buffer capacity of these places is altered by the development.

Wetlands are important habitats for a variety of plants and animals. Research has demonstrated that even small wetlands areas support a surprising abundance of life forms. Large or small, they sustain a broad array of flora and fauna which quite often include threatened and endangered species. Of the more than 200 animals and 100 plants on the endangered list, 45% of the animals and 26% of the plants depend directly or indirectly on wetlands to complete their life cycles successfully. And in the case of plants, there are many more that need federal protection, a good many of which are wetlands dependent or related. All told, 500 species of plants, 190 species of amphibians, and 270 species of birds are estimated to live in this country's wetlands.

While wetlands have this value for us, we can destroy their ability to render these services if we are not careful. Wetlands collect pollutants in the same way that they collect nutrients. Even "biodegradable" ones may accumulate beyond the point where they can be assimilated by the system. Toxic chemicals - not biodegradable - will accumulate in increasing amounts throughout the food web. Once this happens, these scenarios result in occurences like contaminated shellfish and, ultimately, a lack of ecosytem diversity through the death of many species.

Furthermore, although wetlands can absorb a limited amount of pollutants through soil absorption and slow natural decay of contaminants, the capacity to do so can quickly be overtaxed. The more

water level. There are other factors, but the most important one in Louisiana is subsidence - the shrinkage of landforms resulting from rising water levels. Water circulation is also important to the estuarian process- as it is transporter of nutrients and plankton. It also distributes immature stages of fish and shellfish, flushes wastes from animal and plant life, cleans systems of pollutants, controls salinity, mixes water and shifts sediment. Such things as tidal action and wind driven patterns of water movement influence the ecosystem of any wetlands.

In the Barataria Basin, the general flow is north to south. Rain and runoff are the principle sources of freshwater imput. Tidal action permits direct interchange between the estuary and the salt water of the Gulf. One other important factor that assures adequate circulation is the meandering nature of the smaller bodies of water. This helps maximize the interface area and prevents rapid drainage of the systems. Man-built canals superimposed on such natural systems cause serious alterations in these otherwise normal patterns.

Barataria Basin is not a random associaton of organisms but a set of biologically interdependent components, specifically adapted to local physical conditions, and, as such, is divided into five primary and two secondary environmental units. Throughout these units, the food chain is the basic process for recycling nutrients within the system. The various plants and animals are grouped into tropic stages, each stage being one step beyond the primary production level.

Another way the ecosystem can be pictured is as a system of energy flow controlled by photosynthesis. We will learn more about this later. For now, we will divide them into two groups: *primary* - fresh marsh, brackish marsh, salt marsh, offshore areas; and *secondary* - beaches and other elevated areas as chenieres, natural levees and spoil banks.

One thing further, a distinction must be made between wetlands, land that is alternately flooded and drained, and associated bodies of water, primarily inundated areas such as lakes and bayous.

As one of the most highly productive natural areas in the world, the Barataria Basin is composed of a number of sub-areas or environmental units with unique characteristics and independent physical systems. There are five primary units in the Basin: swamp forests, fresh marsh, brackish water marsh, salt water marsh and offshore areas. And, there are two very important secondary units in the Basin: beaches and elevated coastal areas.

SWAMP FORESTS

A swamp is a woody community occuring where the soil is usually saturated or water covered for one or more months of the growing season. Twenty-one percent of the wetlands in the Barataria Basin is swamp forest. The Des Allemands Swamp, located in the northern part of the basin, has the most such area. Some smaller sections of swamp lie seaward, along the fringes of Bayou Lafourche and the Mississippi River leeves. Swamps occupy recent Mississippi River deposits of alluvial soil, while marsh occupies coastal marsh land.

These swamps are largely a detritus-based system with twice as much energy entering through ingestion by detritius-eaters as by herbivores. Energy not consumed within the system is exported to lower reaches of the basin through a network of energy-rich waterways. Swamps have the lowest ratio of water to land. Waterways serve as transporters of material rather than playing a significant role in actual production.

Swamp forests are dominated by bald cypress, tupelo gum, and dummond maple. They have the lowest rate of water -to- land in the basin. One finds crawfish, deer, rabbit, squirrel, small rodents, and seed-eating birds and owls. Other animals include snapping turtles, snakes and alligators.

FRESH WATER MARSH

Freshwater marsh is perhaps the most difficult of the environmental units to define, as it is the most diverse in terms of the number of plant associations. Much of the unit is comprised of floatant marsh, a dense mat of vegetation supported by detrirus several feet thick, held together by a matrix of living roots. Fresh marsh is found primarily between the levee systems of the Mississippi River and the Bayou Lafourche, beginning around Lake Des Allemands and extending seaward to the Industrial Canal. This comprises 223,488 acres, 19% of the basin's wetlands.

The dominant grass is maiden cane followed by spikerush, bulltongue and widgen.

Such fresh water marsh provides an almost ideal growing environment because of its continuous salt-free water supply. Herbivores include nutria and muskrat as well as a wide variey of insects. Carnivores include alligators, and raptors, such as hawks, snakes, mice and racoon. Among the primary consumers are many small crustaceans.

BRACKISH WATER MARSH

Located between fresh water and saltwater marsh, near the Gulf's edge of the basin, brackish water marshland creates an intermediate zone in many ways. In fresh water marsh, rainfall is the main source of new water and it flows only in one direction. Brackish marsh, however, exhibits two-directional water movement because it is strongly affected by tidal action as well as by water movement from fresh water marsh into brackish water marsh. There is a significantly higher proportion of water surface here than in fresh water marsh. As salinity increases the diversity of plant species decreases. The area is dominated by wire grass and salt grass. The Blue Crab is harvested here and brown and white juvenile shrimp both use the marsh as a nursery. Animals include many treasured for their pelts: muskrat, raccoons, mink, nutria, otter, snakes and a limited amount of alligators.

SALT MARSH

The salt marsh region of the basin surrounds Barataria Bay and its interconnecting water bodies. It includes approximately 158,080 acres, or 14% of the total Louisiana wetlands area. Geologicaly, this area is in a transient state, and the boundry between brackish and salt marsh is gradually being moved inland as the coastal zone subsides. The characteristics of the salt marsh are more subject to modification by physical processes - tides, winds, water levels - than are other units. The ratio of wetands-to-water is lowest in this area, and the highly irregular shoreline results in maximum interfacing.

Oyster grass, which has an amazing salt water tolerance, dominates the salt marsh. It plays an extremely important role in the ecystem, serving as a deterent to erosion as well as a nutrient pump, extracting phosphorous from subsurface soils. Other grasses are black rush, and salt grass. Grasshopppers are the only herbivores. The same general species of animals that are found in brackish marsh are found here, except for wading birds which are much more plentiful here than elsewhere.

Bayou Country Ecology

Salt marsh is the main habitat of many of the economically important species of seafood: shrimp, menhaden, oysters, blue crab, redfish, speckled trout, sand trout, and flounder. It is also home to some birds that are either endangered or nearly so. Among them are the Osprey, the Brown Pelican , Peregrine Falcon and the Piping Plover.

OFFSHORE AREAS

As we have already learned water movement both facilitates the delivery of nutrients and flushing away of wastes. We also know that energy not consumed is transported by waterways. The left-over organic material from the production of all four wetland units eventually ends up here.

There have been suggestions that the freshwater discharge may be responsible for the excellent fishing in the areas, since it prevents salt water animals from migration through the river's fresh water.

Let us mention here, in contrast, the "Jubilee" phenomena. This is a large area of oxygen-deficient water, that occurs at times, and causes what is known as a "dead zone." Sometimes dead zones last just a few days; at other times they last much longer. Dead zones cause mass migrating of organisms toward shallower water and, occasionally, fish kills. Their history is very long. They are not new, having been known and noted by fishermen for years and years..

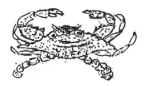

In the Barataria Basin, where tide and river currents meet, rich nutrients circulate. The shallow nature of the area provides maximum exposure to the energy-giving sunlight. In these optimum conditions, the drifting plants of the sea and the rooted plants of the land combine in a virtual explosion of production. No wonder - for this is the heart of the world's largest energy converter.

Secondary environments, or units, found in the Atchafalaya Basin include beaches and elevated coastal areas.

The beaches of the Barataria Basin represent a very small area in proportion to the wetlands. The prime purpose of these beaches is to protect the more vulnerable marsh areas by absorbing wave, storm and tidal energy. These beaches, in general, are drifting in a westerly direction.

There are three basic types of elevated coastal areas: natural levees, cheniere ridges, and spoil beaches.

In the Barataria Basin, the general water flow is from north to south, with precipitation and runoff being the principle sources of freshwater input. Tidal action permits direct interchange between estuaries and the Gulf, particularly at its lower reaches. One important factor in assuring adequate circulation is the meandering nature of the smaller bodies of water within the system. Because the water function of the Barataria Basin is such an intricate and sensitive balance of interrelated factors, man must use caution so as not to upset and permanently cripple this wonderful mesh of ecosystems.

Today, all who love wetlands must realize that love is not enough. So many of the ecosystems of the Earth are in serious trouble. Action is needed, but action without knowledge creates only reaction, and not always of the desired kind. We cannot afford not to know. Both country-wide and globally, people are coming to a realization that we need to learn more about the ecology of our regions so that we can play a part in decisions being made that directly affect our particular areas, their future, and ultimately, our own.

6

Where Sea Meets Shore

*"The rare moment is not the moment
when there is something worth looking at
but the moment when we are capable of seeing."*
-- Joseph Wood Krutch

There is something magnetic about the place where sea meets shore. People seek it almost by instinct, drawn by a force they cannot always define. It is a world where all things seem possible. Yet, a closer look reveals a place of stark reality, for life here at the sea's edge is one of ceaseless activity. Here, where sea and land meet, there is great beauty, mystery, and sudden death. In this world there are only the quick and the dead. Alertness is the name of the game.

There can never be a time when we come to the shore and do not see something in a different light or learn something we never knew before. This is true whether we visit it rarely, or walk it daily, as is my way. The only requisites are eyes that really see and a heart that really cares.

Even now a pair of stalked eyes are watching. The owner, a rather large Sand or Ghost Crab, so named because it is colored so nearly like the sand that it is difficult to see, seems to disappear right in front of our eyes. When the coast is clear, the crab scuttles down the beach, takes a sideways position, grips the sand on the landward side with its legs and waits for the water to reach her. This is a most interesting creature, one in the process of changing from a sea creature to a land creature. But as yet, these crabs are bound to the sea, and in the way described above, must fill their gill chambers periodically. The adult crab can no longer swim for any length of time and does not go into the water further than necessary. As an adult, the crab lives on whatever dune exists on the beach.

Another way these crabs are bound to the sea is that the sea is the place where the female crab must go to liberate her young. By that time, the eggs have already been incubated. She releases the tiny creatures into the gulf waters. For a time they will become part of the plankton and drift helplessly with the currents. As these crab larva grow they shed their cuticle many times; each molt brings them to a new form. By the time they reach their final sea-life size, only a very few of the many originally liberated will have survived. These survivors, called *megalops*, must obey a strange call of instinct. They move

A View from the Heart

Beaches, for the most part, consist mainly of quartz, which is found in almost every type of bedrock and is the most abundant of all minerals. Grand Isle's sand, for instance, is mostly quartz heavily mixed with mud. Any given amount of beach sand may have among its crystals grains of a dozen or more minerals. Although even the hardest and largest rock can eventually be worn away, a grain of sand is almost indestructible. It is the end product - the hard core mineral that remains after years of pressure. There is a certain beauty in that thought. Furthermore, not even the most intense crashing wave can cause one grain of wet crystalline sand to rub against another. Grains of this one type of sand are able to hold by capillary attraction a magical film of water about themselves. Each tiny grain of sand with its covering of water is truly an entire ocean world in which these microscopic creatures, both plant and animal, live out their entire existence. Often people say, "If Grand Isle only had a lovely fine white sand beach it would become a tourist mecca!" In fact, if this were so, the isle would shortly have no tourists, less beach than we have now, and possibly, no land mass to speak of. Why? Weather, winds, and a long shore line, as well as strong currents, do not offer a suitable environment for soft, white, sandy beaches. It would blow away or quickly be washed away.... What Mother Nature gives us is exactly what we need here - a coarse clay sand, generously mixed with mud.

The steed bit his master,
How came this to pass?
He heard the good pastor
Cry, "All flesh is grass."
Anon

What in the world are we doing talking about horses and grass here on the shore. It is not as silly as you might think. In a sense, this Bible-quoting pastor and his steed were both right. From the beginning, it has been the green plant alone that by photosynthesis starts the food web upon which all life on this planet ultimately depends. Once we realize this, in all its implications, we learn to see the green world of plants in an entirely new light. Only the plant world knows the secret of using energy from the sun to create its own food. This capturing of sunlight by plants makes possible all life on earth as we know it. Plants achieve this by a miracle called *photosynthesis* (Greek-meaning "to put together with light"). It is a highly complicated chemical process in which green plants convert solar energy into chemical energy by combining carbon dioxide, water and inorganic salts, with the action of light and chlorophyll, into simple sugars, oxygen and water. Chlorophyll is the green pigment contained in the chloroplast of the plant which acts as a solar energy absorber and is involved in changing and transforming energy until is is finally stored in sugar molecules. Such factors as light intensity, temperature, and availability of water and carbon dioxide effect the rate of productivity, although most plants can continue this process over a wide range of conditions. (**Figure 9**).

Looking out over the Gulf, it may at first appear to be a wasteland. It is difficult to fathom that the sea is more productive than the land in creating, through photosynthesis, an elaborate food web. A cane field or corn field, large, lush and green, seems so much more productive. Yet about 90% of all photosynthesis on Earth takes place in marine plankton and algae, while only about 10% occurs in land plants.

Of what do these "pastures of the sea" consist? They contain both plants and animals, with the plants being far more numerous than the animals. Phytoplankton are the plants; zooplankton, the animals. Microscopic plants are the primary producers in the world. These plants, floating in the upper

Bayou Country Ecology

hundred feet of water, thick as dust in a sunbeam, are efficient beyond belief at the business of capturing light and energy from the sun. Among the microscopic zooplankton, called protozoa, there are innumerable kinds of tiny relatives of crab, shrimp, oysters and jellyfish as well as fingerlings of hundreds of kinds of fish.

Naturally, such pastures extend to marsh and estuary. In the estuary, life is easier and the diet is high in carbohydrates. There is no need for *fast* anything. The zooplankton begin with this diet and, according to species, stay in the inland waters or return to the Gulf. In the Gulf, they will probably, again according to species, turn to predation, no longer feeding only on plants but increasingly on larger and larger zooplankton. The protein of animals now better suits their needs: quick energy, rapid growth, high mobility and rapid reproduction to survive.

Floating communities of plankton drift and spread across the oceans of Earth. These pastures of the sea, invisible to the unaided eye, are unquestionably essential to life in the sea; to all life on Earth. As Jacques Cousteau says: "I assure you our destinies are linked to theirs in the most profound and fundamental ways. All life is interconnected in the great life-giving bank of the sea."

The most essential plant on Earth, constituting six tenths of all plankton material, is a work of art. *Diatom*, from the Greek word meaning "cut in two" are classed among the golden algae. When magnified they have jewel-like beauty. Their intricately sculptured silicon "glass houses" fit together like a box and lid and come in all sorts and sizes. They are clear enough to transmit the sun's rays and strong enough to protect themselves. They are actually practically indestructible. Such structures, millions of years old, have been found as fossils. Diatoms are reinforced with ridges which crisscross in symmetrical geometric patterns so precise they are used to test the accuracy of microscopic lenses. If the pattern appears irregular it is always the microscope that is faulty.

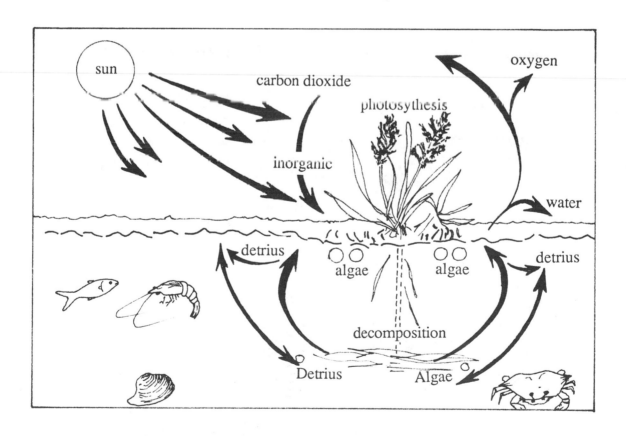

Figure 9

47

The Unseen Sea

Bayou Country Ecology

Another photoplankton of note is the *dinoflagelates* the "whipbearer," which has characteristics of both plant and animal. They may be older than diatoms. Many manufacture their food like plants, others capture their food like animals. They have whip-like appendages that propel them through the water like spinning tops. One of their number, *Gymenecensium brevis*, causes the "red tide" so destructive to sea life. Many dinoflagelates produce light-like fireflies. Watch for them in the Gulf at night.

While the plankton stay in the upper layers of water (the upper 100 to 300 meters of the so-called photic zone) going about their business of photosynthesis, the zooplankton at dawn each day descend to avoid the light of day. Their daily migration is followed by hordes of shrimp and other fish that graze on them. Where heavy concentration of plankton are, marine life abounds. Any shrimper could show you such spots.

Most microscopic zooplankton belong of the family of protozoa - most ancient of all animal groups. The two members of the protozoa of major importance are the radiolara and foraminifra.

The radiolara have their protoplasm contained in a siliceous shell. Foraminifera mainly have calcareous shells. England's famous White Cliffs of Dover are formed of foraminifera shells compacted into limestone of chalk that, through geological changes, reached the surface . Here in our own area, fossil foraminifa are buried deep and are much used by geologists to correlate rock strata for the oil industry. Here they are called *guide fossils*. Some oil-producing zones in Louisiana can be identified by the presence of these fossils.

The smallest free-swimming animal to feed on plankton is the copeopod. Copepods may number up to 3,000 per square foot of sea water. Copepods may eat thousands of diatoms at one time.

Diatoms, little more than a bit of microscopic slime, and copepods, a nearly microscopic "sea flea," what could be less important? Yet, these are two of the Earth's most vital life forms. They are the first step in the food web of the sea - the world's largest habitat/home of four-fifths of all living things. Diatoms, the most numerous plant, perform the miracle of changing lifeless elements into food for countless sea creatures, from the microscopic to the whales. While copepods, the most numerous animal, are efficient gatherers of diatoms. In this way, copepods become the sea's "care package." Eaten by herring, the sea's most numerous fish, the humble copepod begins the food web of the sea.

A great many chemical substances are used in the metabolic conversion of the primary products of photosyntheses to the proteins, lipid and other compounds that make up the substances of living organisms. Most of these chemical substances are present in sea water, but a few of them, such as nitrogen, phosphorous and some growth factors, may be absorbed so completely that they often can be found only in very reduced amounts on the surface.

Within an estuary, or other protected natural enclave where great biodiversity is preserved, a self-sustaining system can be created. The cycle of adaption and competition meet. The various levels of the food chain regulate and support one another. The shallow, well-lit waters become completely self-sustaining.

Unlike the estuary, the open sea, overlies vast, non-productive deep waters with little if any light penetrating to the bottom. While some vertical mixing does occur, it is a function of hydrodynamics and meterological conditions and has not the fully self-sustaining ability of the estuary. This also explains why parts of the sea may be rich in marine life and other parts literally barren.

In temperate regions, in the fall, when the sun's radiation decreases and air temperature drops, the surface of the ocean cools and the action of the wind stirs the water vertically to a considerable depth. This mixing continues throughout the winter and soon the mixing is thorough and the sea is replenished. It would seem this would bring about a sudden growth of plankton, but now another factor enters the picture. Plankton are drifters, and since most can not maneuver on their own, this mixing directs them right into the darker areas. Because there is less solar radiation in winter to encourage growth, little if any growth or reproduction of plankton takes place. In the spring things change vastly. The water stabilizes and thermal stratification begins again, with the warm water now on top and the cooler water confined to the bottom. Now comes an outburst of growth known as "spring bloom" or flowering of the sea. In the early two or three weeks of spring, more reproduction takes place than in an entire year. Whole areas of water are often colored by the bloom. Some mistake this reddish brown color for spilled oil, but it is spring bloom. (**Figure 10**).

In tropical waters, where solar radiation warms the water and establishes a sharp temperature difference (since warm water is less dense), thermal stratification tends to persist. Vertical mixing cannot occur even with the help of high winds. This explains both why tropical seas remain blue and why they have no major fisheries. Clear blue water may be beautiful, but it is not the most productive. In contrast to this beautiful water, our seemingly murky water is extremely productive.

Obviously, all these sea processes must be governed by strict laws of balance, otherwise they simply would not work. Some of these affect adaptation. One such law requires that each species must have its precise ecological niche - location and way of life - within an ecosystem. This rule applies whether on land or sea. Give Nature a niche and soon she will evolve a creature to fill it, for evolution is a complex dynamic process which seeks to fill every available habitat. In the process, new conditions are created that in turn afford place for other kinds of organisms. Creation and occupation of niches is an important part of the work of an ecosystem. In essence, ecosystems are vast networks of interacting organisms and processes that form rhythmical cycles and food webs by which they support themselves and interact with each other, whether on land or in the sea.

Plankton Food by Seasons

One of the most fascinating aspects of the creation of niches is the thousands of ways Nature goes about creating them. One of prominence in the sea is *layering* or *stratification*. There are bottom dwellers, some temporary, others permanent. These are called *bethnic* creatures. Those who spend most of their lives close to the surface are called *pelagic*. Even in these two areas, Nature multiplies the niches by having different creatures of the same area preferring different sources of food. For instance, some predators are pelagic creatures, but most are bethnic. See how it works. The next time you walk through one of Louisiana's lovely forested areas you can easily observe the layering of creatures there, even of the birds. You will find those who nest on the ground, and those who nest only in trees. You will be amazed at all the ways Nature has devised to provide separate niches. Possibly this is due to the fact that another of her laws states that no two species can long occupy the same niche.

Now you can understand how stratification is one method of increasing niches in a given ecosystem. But, this is not a ridged or permanent condition. There are frequently daily and seasonal changes in populations in any given layer due to life cycle changes and other life functions. Nonetheless, this too serves the function of increasing niches and thus the biodiversity of an area. But there are many other ways Nature accomplishes this end!

Time plays a major role in biodiversity. Some creatures hunt or forage only by day, others only by night, still others only in the time between dark and light - twilight or dusk. Many have different seasons for mating. Some have different eating preferences. Others have a preference for one type of soil, water, or bottom material. Bottom material may be mud, sand or gravel. The same goes for preferences for food, and the ability to withstand hardship. The list goes on and on. One interesting example are those creatures of the sea whose active life exists only in the time between the tides, such as the coquina clam and the humpty-dumpty crab.

Let us consider for a few moments the food web of the sea. (**Figure 11**) It is strikingly different from that of the land. Land plants have no direct predator but man, who usually eats the upper part of a plant and does not usually destroy it. But in the sea, the situation is much more complex.

The pastures of the sea are grazed by herbivores animals that in turn are eaten by predators who then feed on each other. In the sea, a state of dynamic equilibrium exists, but it is not simple. Creatures there tend to feed very differently through each stage of their life history. For instance, a codfish egg hatches into larva and begins life by feeding on plankton. Later, it begins to eat small bottom animals and small fish and some larger plankton. However, as a mature adult its diet is mainly fish.

There is a basic "Rule of 10" that states one calorie unit is needed for tissue building for every nine calories of expendable energy. This means it takes most organisms 10 units of food to merely sustain their own body weight. This rule is often referred to as the Law of Diminishing Returns. (**Figure 12**)

Large animals of the sea cannot, for the most part, capture microscopic plankton, for they have no means. (Baleen whales, among all whales, being the noticeable exception.) A series of predators serve to "package" the food into correct size for successively larger animals. This results in a dramatic decline in the amount of energy lost with each transfer, a triangular effect becomes apparent.

We have in the sea, not a food chain but a food *web*, a vast and intertwining web. The word chain has now been replaced by web, since ongoing study has shown that hardly ever is there a single linkage as the term chain implies.

In our eagerness to harvest marine food resources, if we do not understand this feature of the sea's ecology we are likely to over-fish one particular species and thereby overspend Nature's bank account. One simple guideline that avoids doing this is that we harvest as close as possible to the primary level of consumers in the sea's food web. Here the level of reproduction is greatest. Herring are the most plentiful fish in the sea. They belong to the group that are the primary producers of the sea. Also in this group are mullet, menhaden, shrimp and squid. Naturally, the higher the rate of reproduction the less chance of a "wipe-out" caused by over-fishing. Also at this level, there is more chance of success should the time arrive when it becomes necessary to bring any species under artificial culture to maintain the stock.

Some authorities already see this as the wave of the future. At the present, fresh water catfish farming appears to be successful. At a slightly higher level of the marine food web, trout and redfish might eventually offer some possibility for farming, although experimenting done along this line has met with mixed success. The bigger the fish, it seems, the less the chance of success. Shark, for example, is being fished commercially now, but long term, sustained, unrestricted fishing is not going to work.

Thus, if we are to maintain a sustainable catch for today *and* for the future as well, a certain amount of regulation is going to be required. A certain amount of give and take is also going to be required. Everyone needs to understand both the *hows*, and the *whys*, so that rules to safely sustain as much of the gulf resources as possible for the future can be made as fair as possible. Understanding, as always, is the vital ingredient.

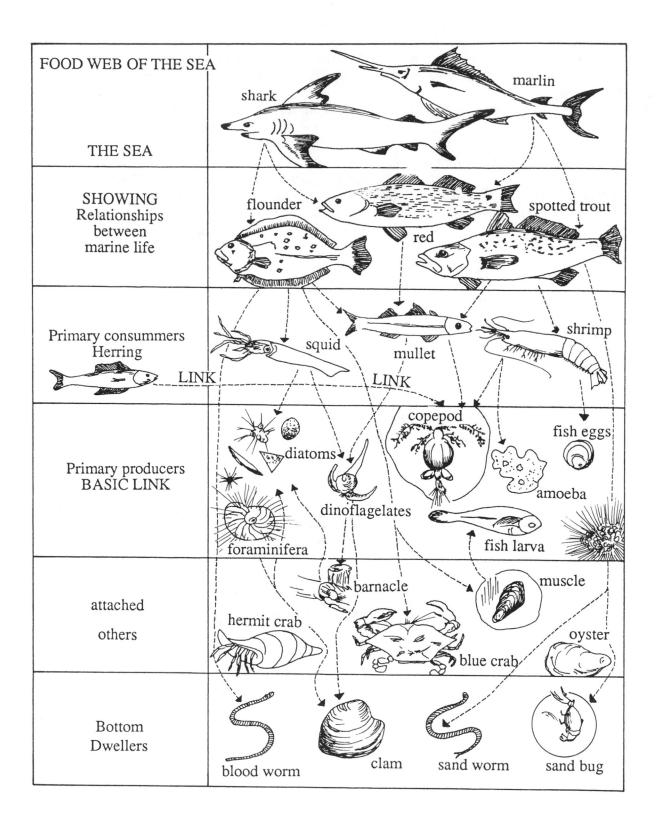

Figure 11

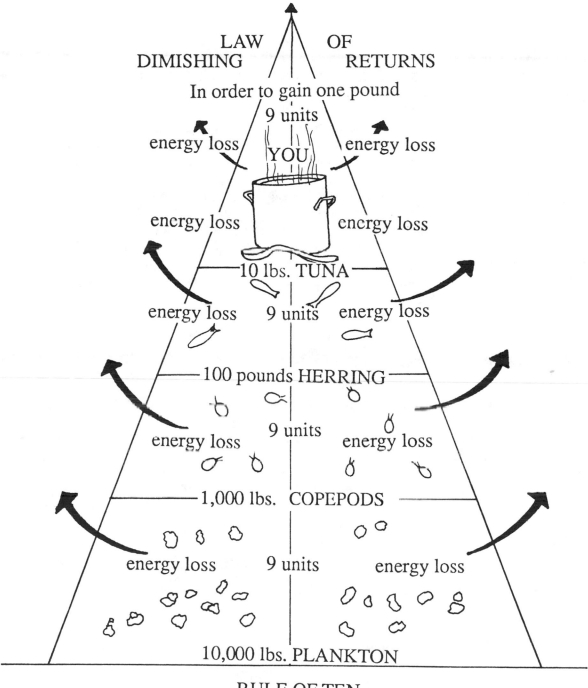

Figure 12

A View from the Heart

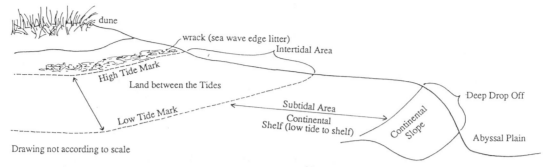

Drawing not according to scale

To be on the beach is to be aware of wind and waves.

As we know, the sun holds the earth in an elliptical orbit that produces the tide and sets the wind in motion. Here we can best illustrate the principle of how the sun effects the wind by referring to the offshore breeze that springs up on warm summer afternoons in South Louisiana, as well as other coastal areas. When the air over the land is heated - and land heats faster than water - it rises. The air over the water, being cooler, does not rise as quickly. This difference allows the air over the water to rush in and fill the vacuum created over the land. Nature abhors a vacuum - hence the breeze.

The wind is Nature's way of maintaining equilibrium in the atmosphere. When the pressure is low and the adjoining air masses are also low, little air can move and the sea is calm. However, if the pressure of the mass is higher, surrounding air immediately begins to move toward the lower pressure, and the speed at which it moves is directly proportionate to the pressure differential. The lower the pressure, the higher the wind. In coastal Louisiana, where hurricanes are a fact of life, this drop in pressure is a sure sign of an impending storm. It is one sign we all know too well.

With the wind comes the waves. Every breeze, or lack of it, is reflected on the face of the water. Let the slightest breeze spring up and the waves reflect it in their motion. Waves are moving bumps and hollows on the water's surface. Such waves are known as *oscillatory* because they are merely an undulation which passes through the water *without* causing the water to move forward - the water simply rises and falls again. It is the motion of the waves that moves forward. The energy of one wave is transferred to the neighboring particles of water and causes another wave, which in turn causes another wave, until a train of waves is produced. Wave motion transfers energy from one place to another without involving an actual transfer of matter. This may seem difficult to understand, but you can easily illustrate it with a rope held between two people. Shake the rope and it moves in waves. The rope itself does not move from the hand, but the waves in the rope do appear to move forward. (**Figure 13)**

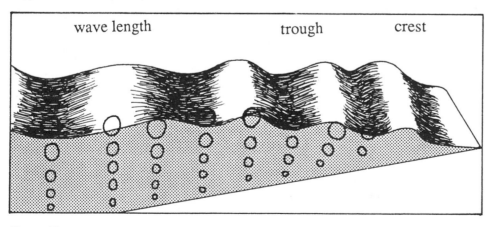

Figure 13

For those who think of waves as being wild and free, it comes as something of a shock to find they are bound by strict laws. Each wave in a wave system has length (distance between crests), height and period (The time between the crest passing the same point.). It also has speed. Only the height factor is independent. The other three characteristics hinge on each other so exactly that any one of the three can be figured by knowing the other two. Furthermore, the relationship between speed and length can be used to show why the wave breaks when it reaches the shore.

54

Bayou Country Ecology

As the wave reaches shallow water, the bottom slows down the wave. How shallow the water must be for this drag to take effect depends on the size of the wave, but when the depth nears one half of the length of the wave, it begins to "feel" the bottom and slows down. As it moves more slowly the wave's length must decrease and, although the wave is now shorter and moving more slowly, it still carries the same amount of energy. Therefore, there is only one thing it can do and that is get higher - thus a plain roller turns into a crashing breaker based on the laws of physics and mathematics, not because it becomes wild or free.

Understanding wave action is important for coastal residents because with such knowledge we can predict what any structural change is likely to do. (**Figure 14**) We need to understand the importance of water depth, prevailing current motion and type of wave action at any given point. We need to study its action at the place we are interested in to determine if it is *cutting* or *building*. When a wave runs into shallow water and breaks near the shore, surf is formed. The water that is thrown forward in the crest of the wave returns as a current along the bottom. This backward current, due to wave and surface currents produced by the wind, is called undertow. When waves reach the shore obliquely, a current along the shore is produced. If you watch the jetties and the like, you can observe how the sand on one side is being *cut away* while on the opposite side it is being *built up* by this action. When any change on a coastline is being planned, *rules* like these, and many other facts about waves and wind, need to be taken into consideration. (Look again at Figure 8 on page 37.)

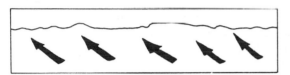

Undeveloped beach currents carry sand along shore simultaneously erroding and building beach.

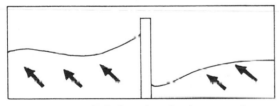

A single groin or breakwater-sand accumulates on upstream side and it is eroded downstream.

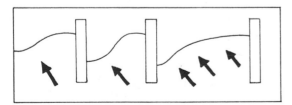

Diagram shows reaction to a multiple groin system.

Figure 14.

The continental shelf begins at the low tide mark and extends out to the continental slope. This shelf is an extension of the land mass built up with the help of rivers that empty into the sea. From the high tide line on out, this mass slopes gently, averaging one foot in five until it reaches its edge which is ordinarily about 600 feet (100 fathoms). Here the angle sharpens to about one foot in ten on the continental slope where it levels out at about the 12,000 foot depth and forms the Abysmal Plain, which has an average depth of 16,000 feet.

The continental shelf represents only about 7% of the total underwater area of the Earth, but about 90% of the world's catch of fish comes from these waters world wide. Most of the offshore oil and gas are presently pumped from beneath the continental shelf. It has an average depth of 470 feet. At its deepest it is 1,200 feet. Its average width is 42 miles. Along the Mississippi Delta, the width varies from 100 to 125 miles off western Louisiana, to 10 miles at the mouth of the Mississippi where it descends to the Mississippi Cone, a formation in the sea floor which is not as deep as the abysmal plain of the Atlantic. The cone is the work of the river and it is cut in several places by deep trenches.

55

A View from the Heart

Beyond the continental shelf lies the mysterious world of the "deep ocean." It is a world of abysmal plains, sea mounts, plateaus, ridges, rises and trenches. While the average depth of the deep ocean is about two and one half miles, its trenches can plunge to a depth of over five miles.

Deep ocean floor deposits differ considerably from those of the continental shelf. Those of the shelf are either extensions of what is found on nearby land masses or of what has been washed out from the land and deposited. Deep ocean deposits derived from the land, in contrast, can only happen where there is no shelf. What we do find is clay formed of volcanic larva decomposed by seawater, and skeletal remains of plants and animals that have drifted down since life began. This forms a sediment thousands of feet thick. Here, undisturbed, is Nature's diary, still neatly arranged, just waiting to be read. A ten-foot long coring tube can give us a 300,000-year swatch of history. Mother Nature's own book, written with her creatures and the material that has drifted down from the surface throughout all the millions of years life has existed on earth. Such material can be read by those with the training to do so. For instance the presence or absence of certain types of shells at various levels in the coring tube speak of the type climate that existed at that time because it is known what type of climate was needed by the various creatures. In another case you might find a layer where a certain creature is very plentiful and another at which it is completely absent. Can you figure what this might tell you if you had not seen it up to a certain layer, or on the other hand if it disappeared and was not seen again in the upper layers of the coring tube.?

An example of the interesting sort of information that has been learned is that the Earth's magnetic field has reversed itself many times over the millennium.

Shoreline fish, as well as brown and white shrimp, are typically estuarian dependent species. They spawn in the Gulf and enter the estuaries to mature. Such species usually grow fast and reach maturity in about a year. Some remain in the estuary returning to the sea only to spawn, while others migrate to the Gulf after their time in the estuary, never to return again. The inshore Gulf and the estuary habitats appear to share a common fauna, but there are certain species which are almost entirely restricted to certain habitats and may be regarded as "indicator" species. An indicator species is an organism, or ecological community so strictly associated with a particular environment condition that its presences is indicative of the existence of those conditions.

Many other factors in the environment affect fish. Perhaps the four most important are depth, bottom type, water temperature, and salinity. These last two need a bit more explaining.

From east of the Mississippi's present delta area to roughly off Pensacola, Florida, the shelf is composed of coarse sand, with many areas of hard bottom, and accumulated shells. It has bays that are clear and salty and support such underwater vegetation as turtle grass. By contrast, most of the Mississippi's water, once free of the delta, drifts westward toward the Barataria Subdelta area bearing part of its load of silt and clay. To the west of the river is an area of mud and sand, finer and darker in color than that east of the river. Underwater vegetation is very rare west of the river. The differences between east and west are interesting both in how they affect the appearance of the shore area and in the fish species present. Shrimp tagged and released by the Wildlife and Fisheries group of Grand Terre, have often been found as far south or west as Texas and Mexico, but they are almost never found east of the river.

All animals have internal fluids more saline than fresh water. Frogs and other freshwater animals do not need waterproof, or gas proof, skins since they breathe through their skins collecting what salt they need and excreting any extra as salt with water in the form of urine. Fish, either freshwater or saltwater, do not have permeable skins and are subject to osmotic pressure from the water in which they swim. Fish gain water and lose salt in freshwater or lose water and gain salt in sea water. Bony fish have reduced kidneys that lessen urine production. The saltwater is reduced in the intestines and salts are excreted through the gills. Some fish are highly restricted as to the amount of fresh or salt water they can tolerate. But the Mummichog (killfish) can regulate its range from fresh to full salt. Sometimes seasonal or weather changes in bays and estuaries make it appear to be less a limiting factor than it is in reality. Often fresh and saltwater fish do indeed mingle. Nevertheless, salinity range is definitely an important factor in the ecology and succession of any given area. In the spring of 1991, this was demonstrated in a most unusual way. Dates were set for the opening of the shrimping season. But because of the unusual amount of fresh water in the estuaries, the result of very heavy rainfall, notice was given that the season could be closed with 72-hour notice if conditions indicated the need to do so. More freshwater could cause the shrimp to move and radically alter the harvest. Nobody, to date, has predicted what will happen, but 1991 will not go on record as one of the better years for shrimping due to the high levels of freshwater in the saltwater estuaries. Nature gives and Nature takes away. The fishing community, for better or for worse, must ride the tide.

We will mention some of the more interesting features of ecological importance of a few fish families found in this area. Notice the ways some of these creatures truly showcase their adaptation to a niche.

Strangers to the area are sometimes confused to see the word "dolphins" included on the fish list, thinking it refers to the toothed whale, *Delphinidae*, which is a protected species. So-called dolphins caught here are a large game fish belonging to the family *Coryaeidae*. It is a favored sports fish, being one of the fastest swimmers among the fish. They are colorful in life and can change coloration. They are caught in both shoreline and open water, depending on the species. As to edibility, this depends on the place and local culture. In Hawaii, the Manimake is in great demand. Here, the demand is quite small. Catching the dolphin is the prime objective.

"Billfish" is a name often unfamiliar. It is a synonym for swordfish, sailfish and marlin. They are a poorly studied group and not too much is known of their history. This seems strange since the family includes some of the most popular sports fish. Sailfish apparently migrate to some extent - their movements correlated to water temperature in the 75 - 80 degree range; while White Marlin seem to like the water a bit warmer and favor areas of upwelling where food is plentiful. They are sometimes found in concentration off the mouth of the Mississippi in mid-summer, but disperse to other parts of the Gulf by late summer.

Tarpon are perhaps the most spectacular of the gamefish. A strong and skillful fighter, he deserves his reputation. They are egg layers and their larval form, like other members of the bonefish family, are elongated and transparent, shrinking dramatically in length at the time of metamorphosis into the juvenile stage.

Members of the mackerel family are migratory fish, but Spanish mackerel can be caught in Louisiana year round, spawning repeatedly from April to September in water shallower than that preferred by the King mackerel. Like the shark, these two mackerel have no swim bladder and must keep moving or sink. There are two additional interesting phenomena to report about them. Kings found schooling around oil platforms are larger in size than anywhere else along the coastline. Also, a two year study showed that 95% of the King mackerel in the Grand Isle area are female. A reason for this has not been found. All Louisiana Kings are larger, on the average, than those off Florida's coast. This also remains a mystery.

A View from the Heart

In summary, how many adaptations to different niches were you able to spot?

Besides making us aware of the tremendous variety, the sea also helps us realize that it offers a showcase of adaptation far greater than the land. Here is a world filled with fantastic adaptation to life's problems. A world alive with color beauty and noise - yes, noise. The first time you hear a caught fish make a strange, loud noise, it gives you a funny feeling, especially those of us who grew up thinking of the sea as the "silent deep."

Of course one usually has to be underwater, or listening to sonar, to hear most of this, but barnacles *crackle*, shrimp *snap*, grunts *grunt*. Drums give out with a raucous *croak*. Saltwater catfish make a sound like the beating of a tom tom underwater; while the Gafftopsail catfish gives more of a sharp *yelp*. Sea robins and toadfish make weird sounds by contracting muscles that vibrate their air bladders. Porpoise have a whole repertoire of sounds. Sturgeon rub their armor like knights of old. Parrot fish sound like kids chewing ice. Whales sing. On shore, fiddler crabs beat their chests and ghost crabs *hiss*; (Wouldn't you just know they would?) In the deep, only the sharks are silent. Why?

Aside from purely communicative, can you see any other possible functions of these sounds? There is little doubt that many sounds are devices to warn off enemies. In spite of the necessity of death to promote life, as we know each creature is provided with help in protecting themselves. The variety of these devices is staggering and vital to the continuation of evolution in the sea. Here, the long picture of evolution unfolds from the tiniest and most primitive creature, to the most highly evolved, with a clarity that cannot be matched on land. Yet our knowledge of the sea, as yet, has many gaps. It is right and fitting that the sea continue to be called "the last frontier on the earth."

Let us remember what Jacques Cousteau tells us: "If man has the privilege to exploit the sea, he also has the duty to protect it."

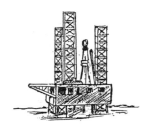

The first offshore oil platform was installed in 1937. Today, there are over 3,600 oil and gas structures off the Louisiana coast. By the year 2000, if current trends continue, approximately 2,400 of these will have disappeared. Hopefully because of a new and perhaps better lifestyle.

A Louisiana Artificial Reef Initiative was launched in 1986 by the Louisiana Department of Wildlife and Fisheries. Under the program finally adopted, oil producers agree to contribute their obsolete platforms and to donate some of their resultant savings into funds to maintain the new reefs. The toppling of the rigs was sanctioned. Now the Artificial Reef Program is administered by the Seafood Division of the Department in cooperation with LSU's Center for Wetland Resources and the Louisiana Geological Survey. One of the best points about this program is that for once we have found something that can be recycled to good advantage. Rigs are now preferred reef material because of their proven ability to provide an ideal fish habitat These gigantic steel structures not only tend to stay where they are better than other forms of artificial reef, but they provide a hard substratum necessary for any marine reef to attract the communities of coral, bryozoan, barnacles, mussels and other organisms requiring a hard surface for attachment. A typical rig, sunk in 115 feet of water, can provide two acres of underwater surface area. A single platform harbors 20 to 50 times *more* fish than the Gulf bottom, according to a study by the National Marine Fishery Service. There are now six such reefs established through the Program, the most recent being in West Delta Block 134, 30 miles southeast of Grand Isle.

Bayou Country Ecology

The 1988 legislative session saw the enactment of the "Redfish bill." It prohibits the commercial harvesting of redfish in Louisiana waters for three years and severely limits the recreational catch as well. The reason given for this was that it was necessary to reduce the harvest because too few fish were escaping the fishing pressure. Redfish were being over-harvested. Why is this happening now?

For one thing, bad days in the oil patch have increased considerably the number of people trying to make a living from commercial fishing. Thus even if the number of fish would remain the same, the catch per vessel will naturally be less. Yet some fishermen, used to a sea of plenty, feel that such regulations are unnecessary: "Mother Nature will balance things out, and provide as she always has." But in truth, this can be true only in a limited sense - Problems do exist that must be solved. The shrimper, the oysterman and the commercial fisherman, are all under severe pressure from regulations at present..

What they need is help. Primarily help in learning and understanding the situation and problems. What they need is basic facts and figures they can understand. They need proof of the situation facing each of their industries if there is ever to be a commitment by all concerned to a better, or at least safe, future for each of these industries.

We also need to realize that these are not problems only of this area. Fishing and fishermen are under pressure worldwide and these are problems that must be solved at all levels if we expect to continue to harvest the sea with impunity and health for both the sea and the fisherman.

"Our troubled oceans belong to all nations,
man's only indivisible heritage besides the atmosphere."
- Dr. Thor Heyerdahl

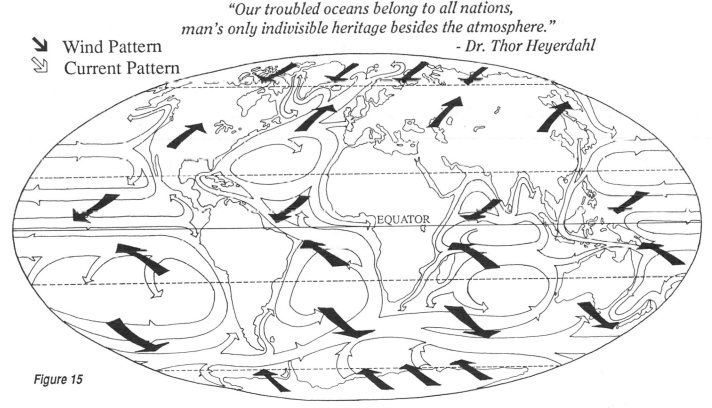

Wind Pattern
Current Pattern

EQUATOR

Figure 15

River and sea may have a love affair, but the sea is married to the wind and together they rule the world. This partnership is so complex and gigantic in magnitude that it might well be considered among the great wonders of the world.

Flow - that one word beyond all others expresses life in the sea. The everlasting flow of tide and current, and the rivers that flow to the sea. Think about how many of the creatures we have learned about spend part or all of their lives as drifters. What a beautiful dynamic tactic for distribution and continuation of life in the sea is this flow and these currents.

The long shore currents carry many creatures in all stages of development to different areas where they find new beginnings, sometimes to areas newly carved by the sea; sometimes to older areas. This living river of creatures comes to seek a foothold, to establish new colonies. Each living creature brings with it the secrets of its long past and an inborn determination for its future. These currents that flow in huge sweeping circular patterns around the oceans of the earth were once considered the longest *rivers* in the world. They are probably, more than any other single element, creators of the marine clime. They are tremendously influenced by the conditions that surround them; sometimes they are warm currents that can change to cold. The greatest of these influences is the wind.

Trade winds are found from latitude 2° - 30° on both sides of the equator. Blowing diagonally toward the equator, northeast by northwest in the Northern Hemisphere, southeast by southwest in the Southern Hemisphere. The trade winds are warm winds showing great dependability of direction. Westerlies are found at 30° - 60° and are cooler and more variable, moving southeast by southwest in the Southern Hemisphere and northwest by northeast in the Northern Hemisphere. (**Figure 15**)

Can't you just visualize these great swirling currents of sea and air? Can't you imagine the tremendous interplay of wind, sea and land? Can't you almost hear the music of this, the "Earth's Symphony."

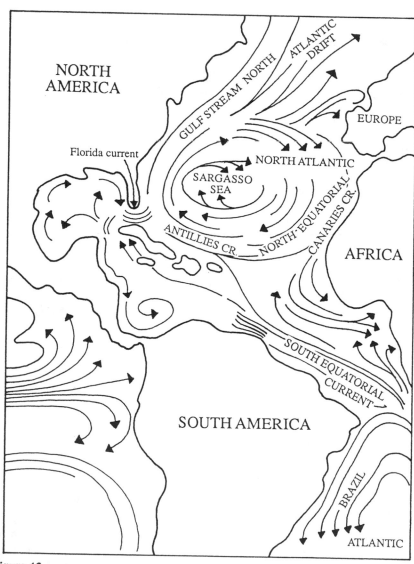

Figure 16

The current that most effects us here is, of course, the Gulf Stream. (**Figure 16**) Many people have the idea that the Gulf Stream is born here in the Gulf of Mexico, but that is not the case.

The northeast and southeast trade winds are the prime movers of the Gulf Stream. Their steady blowing on the surface of the Atlantic causes two separate surface movements traveling from east to west - the North Equatorial Current and the South Equatorial Current. These westward moving currents finally meet the South American continent. Here the South Equatorial Current splits; part of the current flows along Argentina's coast, while a northern branch of the current travels along the northeast coast of South America, to Trinidad. There the South Equatorial Current meets the North Equatorial Current and together they flow past Haiti and Cuba, almost to the Yucatan Peninsula of Mexico. Here the predominately westerly flowing surface current encounters a problem: there is no western

60

outlet from the Gulf of Mexico. The Equatorial currents then are forced to move northward, moving water in the Caribbean Sea through the 50-mile wide Florida Straight between Cuba and Miami. After more maneuvers by the North Equatorial Current and the Antillies Current, this combined flow becomes the Gulf Stream.

In the West Central Gulf anticyclonic eddies, semi-attached or broken off from the loop, at times extend as far north as the Mississippi Delta. Surface layer circulation in the West Central Gulf is dominated by an elongated anticyclonic spiral with the intense flow in the north roughly paralleling the continental slope. Louisiana's long shore current flows in a west-to-east direction - most of the time. When you are tracking a storm in the Gulf, apply this information and see the role currents can play.

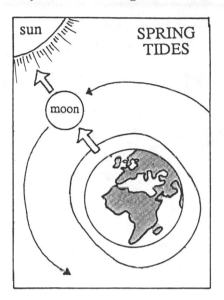

We have talked of tide, but it is interesting to note that in this area tides are among the most minimal anywhere, only showing an average rise of one foot. How strangely this contrasts with Canada's Bay of Fundy - the Cajun home of origin - where 40-to-50 foot tides are normal, and instead of two tides a day, we basically have only one. What is the cause of this? (**Figure 17**)

Tides, like waves are a complicated process. The moon travels around the earth and like the earth it has a gravitational pull. Tides are caused by the pull of the moon. When the Sun and the Moon pull together, high tides are even higher that usual, and low tides, lower. These are call Spring Tides.

At those times when the pull of the sun and the moon is in different directions, the water rises and falls less than usual. These are called Neap Tides.

Furthermore, the magnitude of the tides depend on the depth of the water, slope of the bottom, relation to the ocean, and the width of a channel or pass through which it moves, to mention a few variables. Sometimes some of these factors, particularly the depth of the water and the slope of the bottom, as is the case here on our Louisiana coast, can cause a place to have but one single high tide a day.

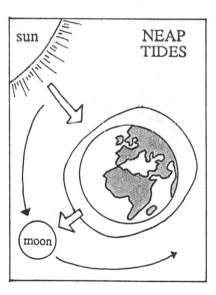

Figure 17

Bayou Country Ecology

Most of this chapter has explored many factors that affect our coast. Now let us pause to sit and watch the birds. No, I am not proposing only bird watching, but that we explore the role birds play in the ecology of this place where sea meets shore. The number and species of birds seen mostly depends on the time of the year. Those you can expect to see here include: Sanderlings, Spotted, Least and Western sandpipers - some of the smaller shore birds known as "peeps". Plovers to be seen here are Wilson, Semipalmated, Black-bellied, including the endangered Piping Plover. The Peregrine Falcon, also on the endangered species list, can be found here, but neither of these are known to breed in the area. Of course, another bird on the endangered list is here - Louisiana's own State Bird- the Brown Pelican. It seems to be thriving here and in spite of some setbacks, is breeding relatively well and making some noticeable progress.

In watching migratory birds, it is fantastic to realize what great travelers they really are. While most follow a standard route, others tend to do their own thing and often turn up about anywhere - Australia, the Pacific Isles or South America.

Shore birds can never again be considered game birds, for at least two very good reasons. First, they all have highly specialized requirements that tend to concentrate them in given areas. Second, they fly in dense, closely bunched flocks, and return again and again to the site of wounded members of their flock. Because of this, they become such easy targets that they offer no true sportsman any real challenge. A single open season on any of their species could spell total disaster for them. Thank goodness for the laws protecting them.

A View from the Heart

While here on the beach, let's take another look at a number of things you have seen and some you have not.

The Portuguese Man of War jelly fish is common, and so is sargassum seaweed. In the summer of 1989, the shores of the Gulf were thick with this seaweed and we were quite literally up to our knees in it. The very lucky have seen a Rare Janthina shell. But what have these three things got in common. They are part of a group known as "Wanderers of the Sea." They live their whole lives far offshore. Only when they are dead or driven in by storms do they ever come to land.

Look closely at the Portuguese Man of War. It is not one animal - but a colony of animals. Each has a job to do - reproduce, find food, protect the colony from enemies. Another colony animal drifter is the Velella, sometimes called the "By-the-Wind Sailor." This is a tiny fellow, a beautiful blue-colored disk about 1 1/2 inches in diameter. It hoists a tiny sail diagonally across its shell and moves into the wind. These tiny creatures live and drift on the sea. Watch for both of these creatures on the beach after a storm. In the Velella's case, be quick; they last a very short time before they seem to melt away - which they really do.

The Rare Janthina, a snail that lives alone in a beautiful purple shell only about 1/2 inch in diameter, drifts on the surface of the open ocean, hanging onto a raft of frothy bubbles. The snail excretes the bubbles which in turn trap air and then harden into a clear surface resembling cellophane. It eats zooplankton and small jellyfish, remaining safe in its camouflaged home that appears to be nothing but a speck of sea froth. The snail wears the soft blue-violet color, favored by so many of the *pelagic* or open sea creatures. This color, when seen from below, where danger could come, makes it impossible to distinguish the snail from the water.... Can you believe how complete are Nature plans, even to this smallest detail?

Sargassum seaweed, another drifter, answers only to the peculiarities of wind and current. Much of it originates in the Sargassum Sea, a relatively calm area in the North Atlantic where there is an abundance of this free-floating algae. A brown algae, with fronds that outwardly resemble stems and leaves, this seaweed has air bladders of varying sizes resembling berries. These are actually flotation devices that keep the algae near the surface of the water. There are more than 150 species of this weed of the sea.

One more familiar creature to consider has been around almost since the beginning of time - the jellyfish, the moon jellyfish, in this case. There is no need to describe him. During spring and summer *Aurelia*, its scientific name, drifts with the current. In summer it sheds its eggs. The tiny larva cling to lobes below the parental mouth, finally breaking away. The young settle to the sea floor, become attached there and grow into small polyps - resembling a plant with fixed base and column-like structure. They winter this way. In early spring, these polyps grow longer and subdivide into thirty or more transverse disks which fall off in succession, each swimming away to become a jellyfish.

Now to the unfamiliar. Last, but not least, we find there is a bug out there - only one. Out of all the thousands of insects known to mankind, there is only one saltwater-inhabiting insect. It is a water strider of the species *Halobates*. His freshwater cousin is sometimes locally called a "Jesus Bug" because it walks on water. Halobates live their entire lives on the surface of the open ocean, riding wave and storm, finding everything they need for life on or immediately below the surface. Stiff water repellent hairs protect its body. It communicates by "calling" - standing and by shaking the surface of the water up and down with the movement of its back and front legs. A middle pair of legs are used to create different "calls" at different rates. Halobates lay their eggs on whatever drifts by - a bird feather, a bit of flotsam.

Bayou Country Ecology

Considering the small amount of dunes along the Gulf shores, there is an amazing variety of plant life. Flowers bloom in their season like bright badges of courage on whatever tiny bit of dune they find. Here one can identify such wildflowers as evening primrose, ragwort, wild hollyhock, Indian potato, blue-eyed grass, dayflower, thistle, false dandelion, clitoria marina, pink star, "rattle bush" and many others.

The Gulf, too, has its own "flower garden" that blooms in the deep waters off the coast. The largest of the coral reefs in the northern Gulf of Mexico lies about 125 miles south of Louisiana Point at Sabine Pass - roughly 90° to 95° Latitude and 27.5° to 28.2° Longitude. Here you can find a twin reef formation truly known as the Flower Garden. Among its attractions is that it houses the most northerly concentration of brain coral in the hemisphere.

As we unwind and begin to think over all we have seen and learned on the beach, one thing above all others lingers in our thoughts - the blind, intense, struggling drive of each creature to survive, to live, to expand, to continue. Yet, at the same time, our senses can quickly be repulsed by the overwhelming disregard for the individual. In the final analysis, this is Nature's way of insuring a huge gene pool from which comes diversity, order, balance and, ultimately, survival of each species.

With the very breath of life, Nature breathes into each creature that tremendous fighting urge to survive, to endure. Then in a final miracle, each and every life form is given that special something that gives it an edge in the moment of crisis. Man's special edge, of course, is his brain power, which can enable him to survive... if he takes the time to use it.

There is a beautiful economy in the environment. It is essential that this be understood if we are to preserve any area, including Louisiana's coast. Understanding how creatures live and interact with their environment is essential to understanding the ecosystems themselves. There is no such thing as an "unimportant" species. Each species of bird, plant, shell creature, or whatever, is important because in the broad picture of life, each ultimately plays a role in the survival of all species, even the human one.

So we learn how the balance of Nature can be so fragile and easily upset, yet, at the same time, strong, aggressive and creative. A force to be reckoned with. A force that despite change, retains the power and ability to survive. It endures not as an individual, but as a complex sum of parts, a living, breathing, dynamic ecosystem upon which we depend and which now, sadly, depends on us. We have, for better or for worse, become a driving force for change, in Nature as well as in ourselves. Can there be any better proof of our desperate need to know?

*"Life would be incomplete without man but
it would also be incomplete without
the smallest microscopic creature."
- John Muir*

7

Marsh and Estuary

"Without the services provided by natural ecosystems, civilization would collapse and human life would not be possible."
- Paul R. Ehrlich

Human beings have always been fascinated by water - but that is not strange. There are those who believe life began in water - in the marsh. Whether this is true or not, there is no question that there are strong echoes of the sea's rhythm in all creatures, man and animal. Our common bond to the sea is that all chemical reactions in man or animal must take place in water that is 0.85% saline solution. Let us search for these rhythms together as we visit the marsh and estuary.

Out here marsh and estuary together continue to work their special magic. If we are watchful, we may catch glimpses of the truth behind the illusions - find answers to questions we, all of us - scientist to youngster - have not learned how to ask. We have come here to search for answers and gold.

Far out in the marsh, away from the activities of the work-a-day world, the marsh is a place of unbelievable tranquility and eternal silence. Once, long ago, silence was a treasured gift of the ancients who believed it the spring well of the mind, the home of the spirit- a revered treasure. How different things are in our world, where the last realm of silence is to be found only in wilderness.

The Sound of Silence

Young people today laugh at the idea of silence being lovely. Noise is their thing, and it fills them; they would have it no other way.

However, there remain among us those who can still thrill to the wonderful quiet peace of wilderness, where the voice of a lone bird seems to fill the world, as opposed to a world where a single voice cannot be heard even when it cries out in desperate need.

Who still listens to the wind, or sings the songs of the seasons? Only the wild ones. Yet, even most people who deny such things, sometimes feel a restlessness they cannot explain. The winged migrants hear and respond. Is theirs, too, a silent call or do they hear a celestial sound we can no longer hear?

Human ears are far from the most efficient of Nature's array of hearing devices. Scientists have begun to suspect, as they probe the mysteries of migration, that some of the higher flying birds can hear echoes of the Atlantic and Pacific surf and this serves to orient their flight. Scientists have found and documented that the Great Whales not only sing what we can hear but they sing also at a level beyond our ability to hear. It makes one wonder how much beautiful, fragile, transient music there is in the world that eludes us completely.

Certain people at certain times may hear such sound and movement unaware. John Muir surely did, and Henry Thoreau too. Louis L'Amour must also have heard it, else he could not have written as he so often did of "spirits that ride the wind" born of this or that special place. And what mariner will not speak of the "spirits of the sea." To hear such music is to hear " the sounds of silence."

"Once, and only once, on the river in the kind of heavy fog seen only on the river - thick, almost liquid, masking everything, creating an awesome silence in which the slightest sound is tremendously amplified -we, too, heard the sound of silence, It happens when one finds oneself listening to the silence between the sounds.

I have a young friend who loves to go into the marsh alone, enjoying the special feeling of elation it gives him. He is touched by its beauty. Unaware he is listening to the sounds of silence racing with the wind, hearing and touching the spirits of this wilderness, sharing with them what Thoreau called "a cordial in sound." He is special. The marsh he loves is special.

The marsh and estuary is a place where one comes to hear the songs of birds and the whispering of wind and water. Yet, its tranquility can suddenly explode in violent storm. It is a place where life and death, beauty and ugliness, walk hand in hand. For all its apparent calm, it is a place of extremes where water remains the thread that binds. Thus wetland and wilderness are truly time binders, tying yesterday to today and today to tomorrow.

What better place to drift and ponder back a thousand years or ahead a thousand years? Either way it is out there, showing us the reality of life bound to life in a line that ties us to the beginning and the end; out there in this trysting place of land, sky and water.

Our sole companion of the moment is a snowy egret - he of the golden slippers, official greeter of the marsh. He sits patiently, as if waiting. I ask him what he thinks of this puzzle called life. He voices his opinion - a single squawk, and flies away, leaving on our minds only the echo of his voice and the shadow of his wings. Egrets never cease to amuse me, always so sober and serious. Did the hand that created them, do so in whimsy?

Here land, bayou, lakes, rivers and swamp seem living veins and arteries of the Earth, yet each is also a separate world teeming with life.

Here in the salt marsh we realize that while the Gulf's regime appears to halt at the waves edge, it does not. Many of its vital processes, as well as many of its creatures, slip silently into the marsh, bays and estuaries.

How is it the same as the Gulf and how is it different? In the salt marsh, the problem of saltwater tolerance becomes even more critical than in the Gulf because creatures born in the Gulf adapt to the salinity of the water, which remains relatively the same. In the salt marsh, however, it is difficult to adapt to a certain level of salinity because fluctuations occur.

Let us look at some of the ways Nature goes about solving the salt problem.

Bayou Country Ecology

Watch the gull, see how he frequently shakes his head. Gulls and some other birds, excrete salt by means of a special gland associated with their noses. Brine blown out with expired air drops from the end of their beaks; their head shaking merely speeds it on its way.

It was once thought that sea turtles cried when they laid their eggs. It was discovered this is merely their way of getting rid of extra salt. On the beach it also serves to keep sand out of their eyes. It would be impossible to name all the ways creatures have to maintain a saline balance; there are probably nearly as many ways as there are creatures.

Do grasses also have this problem? Of course, grasses have a salt problem; all living things do. One dynamic aspect of marine marshland is the changing plant types as we proceed from the shore inland. The basis for the change is the salt content of these waters. The lower and intertidal areas contain high concentration of salt (up to 35 parts per thousand) which limits water absorption in non- halophytic plants, so in this area grow the halophytes- marine plants that have developed specialized structures and techniques to cope with high salt concentration. They are grouped into three categories according to the method used to remove the salt from the internal water of the plant:

1. Succulents - salt uptake compensated for by increased water intake
2. Salt excreting forms - special glands remove salt.
3. Leaf death - salt accumulates in the leaf which dies and falls off at the end of the growing season.

The degree of salinity change any specific species can tolerate varies. For instance, salt grass can tolerate more salt than wire grass.

We must realize, too, that other factors are also important including osmotic pressure, ion uptake, soil aeration and transpiration, among other things.

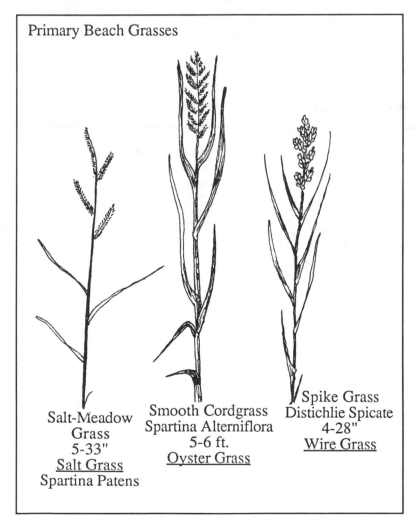

Primary Beach Grasses

Salt-Meadow Grass
5-33"
Salt Grass
Spartina Patens

Smooth Cordgrass
Spartina Alterniflora
5-6 ft.
Oyster Grass

Spike Grass
Distichlie Spicate
4-28"
Wire Grass

Transpiration, or loss of water vapor by plants, is a special problem of salt marsh, since fresh water is not available. To reduce this loss halophytes have evolved leaf surfaces with fewer stomata (openings for passage of air.) In this manner less water is evaporated.

Another problem for the halophytes is seed germination since these plants produce seed that will germinate and grow best in low salinity substrates. So, as you might suspect, many of these germinate in the early spring when abundant rainfall is most likely to occur, resulting in reduced salinity environments.

Nevertheless, most halophytes do not rely solely on this seed method. To increase their number, many have developed underground stems which travel extensive distances, erupting upright shoots at regular intervals - thus insuring a constant supply of water and nutrients from the parent plant.

Another way is that of the mangrove tree. This tree, so tolerant of salt water, remarkably enough has seeds that are not. So, instead of dropping into the water, the leaves remain attached to the parent tree until their roots have developed to a certain stage; then they drop off and drift until the roots touch bottom and anchor in the substrate where the familiar stilt prop roots of the mangrove develop and life begins anew.

It may sound like a ridiculous statement, but marine plants are delicate and hardy at the same time. Hardy in that they have adapted to a harsh environment, but delicate in the fact that these adaptations were achieved only over hundreds of years and any sudden change, with no time for the long process of re-adaptation, can quickly prove fatal. This is but one of the many contributing causes of marshland loss. Something as simple as racing a small boat at high speed through the often very shallow and narrow *tranasses*, or trails, in the marsh can damage the upper roots of marsh grass by loosening the soil. People who truly love the marsh often do this for fun and don't realize the damage it does.

Grasses consist mostly of cellulose, a substance that strengthens plant walls. Since it is quite resistant to digestion by most animals, there has to be another way for its locked-in energy to become available to other life forms. When the outer leaves fall or when the whole above-ground portion of the plant dies in the fall, at seed formation, or from freezing, trampling, or washouts, whatever the reason, once the plant hits the ground bacteria immediately begin the process of decomposition. This reduces the dead plant to detrius. Bacteria are the most primitive form of non-green, single-celled, microscopic plants. Bacterial action breaks down the complex molecules locked in the dead plants and animals and changes them into simpler molecules that are

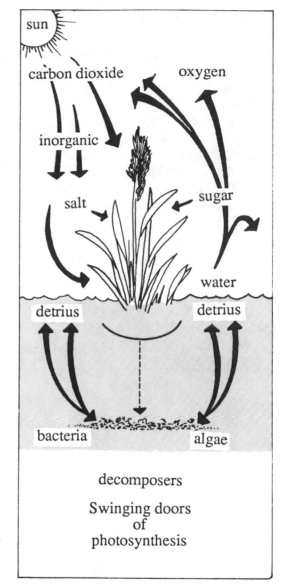

Figure 18

usable by other plants and animals. These chemicals released from the dead plant and animal material by bacterial action become available again and again. This is an extremely vital process because carbon and nitrogen, so necessary to life, occur in available forms only in very limited amounts. Bacterial action is literally the flip side of the cycle of photosynthesis. Indeed, bacteria are literally the "swinging doors" between the organic and the inorganic world. They shuttle marsh matter through chemical pathways at whirligig speed. Another look at **Figure 18** will help you better see this.

Some bacteria are colorless and for the most part unable to manufacture their own food or even to take in food and digest it. Digestion is performed on the outside, which results in food breaking into smaller and smaller pieces as the process continues. As more plant material is digested by these micro-organisms, the end product, detrius, is more nourishing than the original plant itself. Small particles and their associated bacteria are spread about the surface of the marsh by tidal action. This is the start of the salty soup of the marsh, a "gumbo" rich in nourishing broth which is the first food for most estuarine life. As a result of such bacterial action there are about five tons of suitable detrius and algae food available each year for every acre of marsh. Detrius generally refers to all organic matter involved in the decomposition of dead organisms.

Bayou Country Ecology

In Louisiana, the bacteria and algae combination that constitutes most detrius better than doubles the food value of the original living plant. These algae, and the detrius with which they are always mixed, are the most abundant food of the marsh. Once in the "gumbo," these algae use some for producing their own food by photosynthesis, and in turn are then fed on by slightly larger one celled animals and fingerlings of hundreds of kinds of fish as well as the tiny relatives of the crab, shrimp, oyster, jellyfish and others.

There are forms of algae that live on the surface of the mud, migrating up and down, adjusting conditions of light and heat to their own advantage as the tide covers and uncovers their habitat. There are also floating algae that in some places enter the marsh at high tide and leave at low. The many animals that feed on this detrius and algae mix can be identified by the presence of tentacles, filter, sieves, scoops and other such feeding devices. Those feeders include clams, oysters, mud snails, shrimp, fiddler crabs and mullet. To what group do these algae belong? They are part of the phytoplankton and, as such, begin the food web of the marshland.

The next level on the food web belongs to the insects. Apparently only limited study has been done on the role of insects in Louisiana marsh ecosystems. But enough has been done to establish the fact that they have a very profound affect by sheer virtue of their numbers, both as individuals of a species as well as in number of species. About 1,000,000 different species of insects have now been identified. However some estimate that the final total will be as high as 30,000,000 species. This would then represent well over three-fourths of the entire animal world.

Insects have no salt problem. Their outer skeletons provide a waterproof covering. They breath through tiny tubes whose passages carry air to all parts of the body. Only the very tips of their chiton or armored plates are not waterproof and those sometimes have valves that can be closed. What an amazing world is the Lilliputian world of the insect.

It is hard to realize that a creature not much bigger than the period at the end of this sentence can have the capacity for walking, flying, working, mating, not to mention being able to deliver a sharp bite.

In Louisiana marshes, the insect's prime role is in the food web. Plant bugs, plant hoppers and some beetles feed on plant exudate. Tiny grass flies live within the stems of some grasses, eating plant tissue. Grasshoppers and other insects eat Spartina or other grasses, and, in this way, aid in the transformation of plant fiber to detrius. Other detrius feeders that need to be mentioned include turtles, herring gulls, various egrets and herons, as well as the larva of some insects, including mosquitoes.

In turn many insects feed on the mosquito in its various stages. Mosquito larva and pupa live in the water and are eaten by fish, especially the top-minnow (*Gambasis affinis*) known as the mosquito fish. Mosquito larva is also eaten by different flies *(Tabanus)*, among them the horse fly, deer fly, sweat fly and picture-winged fly. While no one animal depends solely on a mosquito diet, many insects, like dragon flies, wasps and predatory flies, as well as birds, bats, and fish do eat large amounts of mosquito larva and adult mosquitoes. The smaller predators of this group are, in turn, eaten by larger animals.

So we find insects close to the primary level of the food web, helping produce food and serving as food. We also learn that insects, as do all other animals, have their herbivores, carnivores, predators and preys.

All flies are members of the *Diptera* family, of which there are 80,000 known species. In Louisiana, there are 57 or more species of mosquito. Actually, 57 is a relatively small number when you realize there are 3,000 species known world wide. The most infamous of all in this coastal area is the salt-marsh mosquito (*Andes sollicitans*) who lays its eggs in the ground at the edge of salt and brackish marshes. The eggs can remain viable for several months, surviving even during dry periods, only to hatch within 30 minutes after being covered by rain or a high tide. In seven to ten days, the lucky ones are ready to go to work on their unlucky victims. Females are vicious and, unlike many other mosquitoes, do not attack only at dawn and dusk, but any time at all. Their life spans are longer than that of the average mosquito.

Besides size, what else makes the insect world special? Many things about insects make them important, not only to marsh ecology but to everywhere else as well. A world without insects would be a world without clover, alfalfa, apples, less oranges, fewer garden vegetables and no honey. There would be fewer birds, fewer fish, fewer flowers, fewer animals. This is due, in part, to the role insects play as pollinators. Yet, there is more. Much wilderness vegetation, important to soil formation and control of soil erosion, depends on insects, both as pollinators and as agents of decay. To some extent, by feeding on organic matter, they also play a role in limiting waterway pollution.

The insect world is an endlessly adaptable, enormously successful, tremendously varied world full of surprises. Here are a few examples: The flight of many insects is truly amazing. Each of the dragonfly's wings is separately controlled. Their owner can hover, rise, descend, move forward or backward at any angle. The deer bot fly is said to be the earth's fastest flying creature. Insect reproduction rate is incredible. For instance, the turnip aphid of the Gulf Coast region can produce

Bayou Country Ecology

35 to 46 generations annually with each female, and there are no males, giving birth to 50 to 100 live young. Some insects can go up to a year without eating at all. Others are capable of walking around for days without a head. This, in part, is because they are different from all other living animals.

The insect world is a world of almost total programming (instinctive or inherited behavior). The rest of the animal kingdom depends far less on programming, far more on individual learning and adaptive changes in individuals. The insects are in a very real sense, tiny robots whose activities respond only to a minor degree on reaction to current stimulus and to a major degree on genetic programming that must have taken place millions and millions of years ago. But there is always something new to discover. Renewed interest in studies of the common dragon fly, for example, are focusing on how it flies the way it does. The areodynamics involved in his flying technique is far superior to that of our most advanced helecopters. The experts are still trying to learn the secret from this insect that has been on the Earth since before the age of the dinosaur.

Insects offer man both his chief competition and his greatest help. They eat everything, our food, clothes, houses, plants, animals and even us. Yet, as in all Nature, there is balance. The insect kingdom represents a balance wheel, tiny, intricate and very necessary.

Next in the food web of the marsh come the birds. Here, as in all other forms of life in such an ecosystem, diversity is tremendous. Among the smaller permanent residents in coastal Louisiana are the seaside sparrow, killdeer, belted kingfisher, boat-tailed grackle, red-winged blackbird. The seaside sparrow is fairly representative of the group who prefer life near the ground, scuttling between the grasses. Its most striking feature is its bill, which has adapted to accommodate its diet of snails, crustaceans, and other such shelled creatures. Winter finds other birds of this group here, as the swamp sparrow, sharp-tailed sparrow, and marsh hawk, to name a few.

This grassy habitat is preferred by other groups of birds, too. Some of the more or less duck-like birds are the gallinule, rails, sora, coot, and grebe. On the bayou, clappers are a favorite game bird, here called the "marsh hen." They are seldom found away from the salt or brackish marsh. Being secretive creatures, they are better located by their call than by sightings.

Next are the wading birds, who occupy a different niche. The permanent resident herons include the great blue heron, the Louisiana heron, the great egret, the snowy egret, and the black-crowned night heron. Winter wading birds include the reddish egret, the little green heron, and the yellow-crowned night heron. Only a few migratory night herons winter in Louisiana; yet, as a summer resident, it is more common than the black-crowned heron. The bittern is a bird who only summers here.

Gulls and terns, while not confined to it, are very much part of the marsh and estuary. Where they are found depends very much on their species. Only a few of the many gull species are common here and only one is considered a permanent resident, breeding here. Those found here include the herring gull (the most common gull anywhere), the ring-billed gull and the laughing gull. While the first two are more or less scavengers, the laughing gull is more of a fish eater, liking to soar on thermal winds and often seen in open water. Its numbers seem to stay steady and plentiful here year round, and it is definitely known to breed here.

There are more species of terns here in this coastal area than gulls. Four are permanent residents: the royal, foster, sandwich and caspian tern. Terns have forked tails and are somewhat smaller and more graceful in flight than the gulls.

One more bird - a personal favorite - seen in the marsh more often than on the beach, is a "bec á ciseaux" - the black skimmer. This most unusual bird feeds by skimming low over the water with the long lower bill, skimming just below the surface of the water. Like a miniature fighter plane, he appears to strafe his target with wings beating furiously. Other times he simply glides. He does his hunting mainly at sundown. He is the only bird who has vertically split eye pupils like a cat. (Did you notice the time factor as a niche here, plus the special eye adaptations?)

Another interesting adaptation is seen in bird eggs. They display the complexities and infinite detail of Nature's ways. Rails lay eggs that can survive in saltwater; wrens do not. The size of an egg is not indicative of the size of the parent, but more indicative of the stress the young are going to face. The little life is further developed in the bigger egg and some have down or soft feathers when hatched. Killdeer and sandpipers and some others hatch out birds that can run a few hours after hatching. Ducklings can take to the water when only a day old. Ground nesting birds are subject to more predators than bush or tree nesting birds, but, in contrast, the young of bush or tree nesting birds are often blind, helpless and naked when born, and their parents must care for them for several days or more.

Very important in the food web of the estuaries is the role of the primary producers. They are the foundation of the nursery grounds of the next generation of fish. The estuaries of the marsh, however, bear little resemblance to a "cradle" as such, but that is exactly what they are. Here beneath a shallow blue-green covering of water and a soft rich mud bottom mattress, thousands of tiny wriggling forms that will grow into the next generation of fish are here, either born here or borne here by gulf currents that carry them in from offshore spawning grounds. In Louisiana estuaries, perhaps more than 50% of our country's future sea food are dining and growing fat on Nature's own version of "Cajun Gumbo". This analogy is not too far off, for the vast shallow nature of the estuary provides maximum exposure to energy-giving sunlight, allowing the rooted plants of the land and the drifting plants of the sea to combine in a virtual explosion of activity. Detrius and algae are mixed, blended, warmed, stirred and served - the nutritious, abundant soup-like food of the marsh.

It is estimated that such coastal estuaries can produce ten tons of plant material per acre per year. This can compare to one and one half tons of wheat produced per acre on the world average.

Besides these fish, there are many other creatures for whom the marsh is more or less a permanent home. Among the group are worms of many sorts, snails, and other shelled creatures, jellyfish, tiny relatives of the shrimp, crabs, and small fish.

The minnow types are plentiful - sheepshead, marsh killifish, gulf killifish, salt-marsh top minnow,

Bayou Country Ecology

bayou killifish, longnosed killifish, mosquito fish and sailfish molly, who also ventures into protected higher salinity water and who bears live young. There are many other kinds of small fish to be found. They must play an important role in the food web, the number of kinds here attests to that.

One of the larger fish common to the marsh, as well as the shallow waters of the Gulf, is the gafftopsail cat. Considered to be a nuisance fish, they are, nonetheless, interesting due to a strange adaptation: the male takes the eggs the female has laid into his mouth and carries them until they hatch, never eating or biting on a hook during this time.

The food web of the marsh does not stop at the water's edge. This "trembling prairie" has among its grasses and along its water's edge, many forms of life. Spiders are here - wolf, crab, jumping, as well as the large and beautiful golden-silk spider. Here also are the "tullaloos" (phonetic spelling, since no one seems to know how to spell it). These little fellows are, of course, the comical fiddler crabs. Hermit crabs are here, too. They live by moving to larger and larger abandoned shells as they grow. You can find them in all kinds of shells, especially moon and oyster drill shells.

One of the most delightful and interesting creatures of this area is the very abundant porpoise. Those in this area belong to the family of the Atlantic bottle-nosed dolphin. They are found both in the Gulf and in the "inside waters." They are heavily concentrated around Barataria Pass. Fishermen and porpoise share one hatred in common - sharks. Fishermen will often tell of seeing porpoises kill sharks. This is correct. Remember how we talked about nature giving each creature its edge. Nature has given some shark species up to twenty-seven rows of teeth that serve as one of the most effective cutting devices known to exist. Yet, because they have no bony structure, only the pressure of the water helps support their internal organs. If removed from the water they easily die of damaged or ruptured organs. The same thing happens when they are battered or banged about in the water. This is easy with the team effort that is natural to these highly intelligent porpoise. It seems the hand that gives also takes away. Nature may not be neat, but she certainly is efficient.

A View from the Heart

"The instinct to protect the next generation drives some automatic motor response in the dolphins and other species. To me this is marvelous because the successful replication of life is what makes our Oasis in space such a rich biomass, fecund and prolific, forever generating and nurturing new organisms."
- Jacques Cousteau

We are finding that plants and animals by gathering in complex ecosystems, contribute a great deal to the creation of their habitat. All such communities are important. Such an ecosystem community is that of the oyster; this community is nearly completely built around them.

An oyster has gills - numerous, small, beating, hair-like cilia that draw in water and expel it from between the two valves of its shell. Not only do the gills serve as a breathing organ by pumping the water in and out, but also collect food on their surface and carry it to the mouth, transport and excrete the waste. Thus, the oyster strains plankton and detrius from the water, consumes some, but rejects much of it too. The rich concentrated nutrients, algae and microscopic creatures, already processed fall back out of the shell, Thus it supports a multitude of other filter feeders who live near the oyster to take advantage of his "wasteful" ways. Among them are many varieties of small crabs and worms.

Surface attachment is a great attraction. It attracts another group of worms, barnacles, and at least one species of small fish. In other parts of the country the oyster has more competition for attachment from such things as mussels, slipper shells, and the boring sponge, Cliona. Here in Louisiana this the Cliona group presents no real problem, although I have seen some oyster shells showing Cliona's art work.

Fish are attracted to this community. "Fishing is always good around an oyster reef."

For years the toadfish was considered a predator of the oyster; in fact it was called the "oyster cracker." Fairly recent studies have indicated he is indeed a friend who chews instead on the oyster's enemies. Another friend is an odd little fish named the naked goby. This little one-inch fellow is almost completely lacking in scales. He has a curious formation of fins on his underside that act like a suction cup which he can use to hold on to any hard surface and not be washed away. His diet consists of worms and small crustaceans and, in this way, really benefits the oyster community.

Black drum are here. They are vicious predators of the oyster. With a powerful structure in his throat, he grinds up the oyster shell. He can work havoc in a short time. The blue crab is also considered a pest to the oyster. But, the oyster drill is by far the worst. This gastropod, a member of the snail family, has changed very little over millions of years and can take a heavy toll in an oyster bed. But enter the Cajun! He eats them. Known locally as "biganos," the oyster drill is the Cajun version of the more expensive escargots. The moon shell is also a predator of the oyster. One creature occasionally found

Bayou Country Ecology

in oyster shells is an extremely tiny crab who lives there feeding on everything the oyster does.

So we see in this community an intricately woven fabric of life, with each animal linked to the other and its surroundings in a vast arrangement of relationships. It is a place of conflict, compromise, accommodation and stability... stability that in the final analysis depends on the oyster.

Perhaps each of us has found answers, but have we found gold? Certainly. It is right here where it has always been. Fields of marsh grass, "flotant" islands of grass, chenieres rich with live oaks, still waters thick with detrius - viewing the estuary in this way it is not hard to see the "gold" we have here in Louisiana. More precious than "black gold" because it is a renewable resource, we have discovered the lovely green gold of bayou country. When the days of our oil patch end, green gold will still be here, just as it always has, ready to sustain us, if we sustain it. In order to do this, we must do as we have done today. We need to listen to the marsh, we need to understand its needs, for they are ultimately our needs; its well being, ultimately our well being.

But now the day grows late. Floodtides set our green world afloat between heaven and earth. Evening suffuses the sky with glowing colors. The water captures color and turns it into undulating beads of many colors. As we wend our way home, riding a rail, beauty below, beauty above, our minds can wing out on a flight of imagination. Has anyone seen an estuary - that's what we are looking for now . . .

Symphony of the Marsh

It is late afternoon. A silver seaplane roars along the bayou toward the east and takes off. Hardly has quiet returned before two porpoises swim by. Moving like gently rolling waves, they seem to be tailor-tacking the water. They are heading west along the bayou.

A trim shrimp boat comes up the bayou and slips gracefully into the fishing dock. We hear the skipper talking to a man on the dock. His heavy, yet soft Cajun accent reminds us this is a special place. This is South Louisiana.

The tide is high. The salt water of the Gulf and the fresher water of the bayou mingle. Two fish, one from the Gulf and one from the bayou, swim side by side. A mullet jumps horizontally, clearing two feet of water. It returns to the water with an echoing slap. Along the bayou's edge three brown pelicans skim low over the water.

The sun begins to approach its mooring. Another porpoise, this time a larger one, makes its way west on the bayou. Let us travel with him, in our mind, beyond the places of commerce and oil that mark the inshore area of the bayou. A helicopter passes low overhead on its way to a spot a child once described as "where the copters go night-night." Boats pass us going both ways, some large, some small - work boats, bent on one errand or another. Some bring workers home from offshore rigs. Fishing boats are homeward bound with the day's catch. Pleasure boats are here, too, one tows a skier whose wake catches the colors of the setting sun. Strangely enough our porpoise friend does not seem bothered by all this and we, with him, glide peacefully on.

Finally the shoreline bustle gives way and we move further out into the bay. We are going just where we had hoped to go: out of the bay and beyond into brackish marshlands. The quiet is so thick you would cut it with a word, and we find ourselves whispering so as not to disturb this brooding silence. Mosquitos in swarms, like grey and torn curtains, brush by. We step off onto a dock to enjoy the night and its symphony. As we watch, the night comes with a velvet touch, a touch that like the tap of a conductor's baton, and thus begins a strange symphony, "A Night on the Marsh."

The overture is the universally familiar "Serenade for a Summer's Night," but soon we are aware of new sounds, new instruments, new singers. All about us the drone of mosquitos is punctuated by an occasional buzz in the ear. At our feet, the staccato snapping of the "tulla-loos." From the water comes the powerful beating sound of the black drum fish. For added effect, a chorus of unknown voices rise from the water. From time to time, the cutting sound of wings and the buzzy "pe-ent" of the night hawk pierces the sky. We hear too the barking "quawk" of a black-crowned night heron. From the distance, a swish and puff as a porpoise breaks the surface, up for a breath.

The moon is a silver boat tonight but its light is enough to kindle the waters own lights-the dinoflagellates, marking all that passes with an eerie halo. The comb jellies, thick as upside down fireflies, twinkle in their own and the moon's reflected light. We drift and dream in a balmy gentle darkness where time has lost its meaning. So it comes as no surprize that in the darkest dark, that is just before dawn, we see rolling across the marsh *feu-follet*, or "false fire" of Cajun legend - Louisiana cousin to the "Will-o'-the-wisp." We are tempted to follow, wondering if it will lead us out of the marsh, or further in, to lose us forever. It depends, *mon amie*, it depends...

Slowly dawn begins to lift the veil of night. The music becomes more *dolce* then *andante* but does not stop. Now instead of a symphony of sound we are experiencing one of color and light. Here on the delta, with water on all sides, the clouds and the colors of the sky are among the most beautiful in the world - a bit of *lagniappe*, it seems, for having to tolerate such high humidity. Where better to see the dawn than where the sky touches earth and water in every direction. We watch the swiftly moving clouds change color over and over in a tone poem. Finally we literally hear a thunder of color as the *crescendo* and *finale* are reached with the appearance of the sun.

Suddenly it is the silence that is deafening. Slowly, heart beat-by -heart beat, we become aware of the earth and the fresh beauty around us as life stirs to meet a new day.

The third movement begins, The Dance of the Herons. Wading birds wake early and hungry. Like any dedicated fisherman, each species has its own technique for catching fish. The great white heron simply stands and waits in a somewhat awkward position untill a fish swims by and then he snaps it up. The snowy egret adds a shuffle, or perhaps a moon walk. With wings outstretched it stirs the water with its "golden slippers" to flush out its quarry. The Louisiana heron, as you might guess, tops them all with what might be called a *pirouette*. Stepping forward, wings extended, she turns in place, raises one wing and puts her head beneath it to gaze into the water. Then completing her turn she raises the other wing and takes a look around. The slow erratic shadow she casts helps lure prey into her trap.

The raucous chattering hak! hak! hak! of the marsh hen sounds nearby and the bird makes its appearance between two closely spaced clumps of grass at water's edge. We are amazed at its size; so slim it has passed between these two clumps without even so much as stirring a single blade. Then we remember this is the origin of the expression "skinny as a rail." Rails have the ability to compress their body laterally to keep from moving grass which would alert enemies to their presence. Marsh-hen or clapper rail - one and the same - for him it is bath time.

Bayou Country Ecology

A bush near to where the rail is bathing is literally white with blooms, about it hundreds of yellow butterflies flutter. Fecundity and profusion are the way of the marsh and estuary, the way of all Nature, for that matter.

At times the sky is nearly black with birds in passage. Yet now, high overhead, a single gull laughs as it soars on a thermal updraft. It sails rather like a hawk, but its laugh tells us it's a gull. Less of a scavenger than many gulls, this laughing gull is a true fisherman in a fisherman's paradise.

Where does this marsh stop and the estuary begin? The whole subject of estuaries is fraught with irony, beginning with the simple fact that we cannot even adequately define them. They slip through our attempts like rain through trees. We cannot hold them. They are places of blurred edges, undefined lines, indefinite boundaries. The dictionary is no help - defining estuary as a tidal bay formed by the submergence of the lower part of a river valley. Doing only a bit better, a professor defines it as a semi-enclosed coastal body of water with free connection to the sea. A coastal resource specialist, justifiably, threw the whole thing up in the air and came up with the idea that it is more a state of mind than a physical reality. A writer in **Wilderness Magazine** arrived at perhaps the best possible definition. They are pathways, switch-yards in the global nutrient cycle, importers of raw material, exporters of food stuffs. Perhaps, metaphorically, we could consider marsh and estuary Siamese twins - two yet one. But perhaps there is more. . .

Some Louisiana people for whom such a place has been home for generations, could have told you long ago what researches have learned but relatively recently. They would have told you "food stuff"- nutrients, are concentrated in many parts of the inside waters in amounts many times those found in other parts, or even in the open Gulf. They knew too, that these places trapped sediment and that they contained abundant life of all sorts. Such old timers could not have known how effective these places were in filtering out petroleum waste, or pesticides, nor did they know that marsh and estuary were a powerful biological engine driven by wind, tides and current, powered by the sun. But in their own way they sensed it, and respected such places. They knew the rhythm of the tide. They knew the significance of the water colors. They could *read* the water. In much of their knowledge there was habit, five senses, and perhaps one more. Great love can sometime give us that sixth sense about the welfare of that we love.

Having captured all we can hold for now, let us catch the next porpoise "shuttle bus" and begin our return to reality. We meet reality as we pass an oyster boat on its way in with the day's harvest. A boat load of sports fishermen, hopeful of collecting their share of marsh bounty, speeds by. We pass an anchored boat from the Louisiana Wildlife and Fisheries. Agents are working with instruments and notepads - a small laboratory in the marsh.

What have we found the marsh and estuary to be - a harbor, a place of commerce, a scientific laboratory, a cradle, a bird, a fish, an oyster, a gigantic energy convertor, a fisherman, a hunter, a switch-yard in the global cycle, an intricate web of life? Above all, we find it a place where earth's long past, as well as its present and future, are alive and well and overflowing with bounty, beauty, and immense promise.

How can we measure or describe the importance of such a place? How can we let it remain in danger?

Thoreau was right; wilderness is the source of Nature's strength. One can feel it in the vast, calm, measureless days spent here; swallowed in space, swamped by beauty, afloat in time. In wilderness, the age-old cycles of life and season are alive and well. Its purpose now, as in the beginning, is to simply be. Nothing more, nothing less. Doesn't your heart cry out as does mine: "Wound it no more!. . .Let it be!"

A View from the Heart

80

8

A Journey Toward Understanding

"We all travel the Milky Way together, trees and men."
- John Muir

We live on a top, not a child's toy, although like a child's toy it can easily be destroyed. Earth is a lovely, pale blue top that can be seen from outer space spinning on its axis, completing one revolution every twenty-four hours. Like any top, it tilts as it spins and our seasons change . Once every 365 and one-fourth days it completes one journey around the sun. Once in 200 million years, it completes one journey around the Milky Way

The surface crust of Earth floats on the mantle. The mantle's crust is made up of large slabs called "plates." These plates fit together, though sometimes they shift, causing earthquakes. Continents, too, are always moving slowly. Presently, North America and South America are drifting slowly west, while Europe and Asia are drifting east. Africa may eventually split apart. New volcanic sea isles occasionally appear, while other ones occasionally disappear.

As if this were not enough, shore lines, seas, and currents are constantly changing, ever on the move. Water is never still. Even our air is never still, nor the blood in the veins of all living creatures.

We cannot think, cannot plan ecologically unless we remember and make allowances for the fact that we live on a flying top and that perpetual motion is the only game in town.

South Louisiana is at the heart of the Earth's largest energy converter. But do we understand what this means? An energy converter simply changes one form of basic energy into another. This happens in an ecosystem. Plants, for example, are energy converters. They create the basic energy for all other life.

As it concerns us here, energy is the dynamic flow of power and material in a given geological environment. All adaptations made by individuals and species are also forms of energy. Why do we need concern ourselves with this?

Human beings and human activities are imbedded in and dependent upon the physical systems that support their lives, according to Paul Ehrlich. Wendell Berry says "We will ultimately fail at our attempts to protect and preserve our wilderness if we do not preserve the ecological processes that bind them together." Energy is what runs these systems. Knowing the role of energy is critical to understanding ecosystems.

The key aspect of the behavior of energy in ecosystems is contained in two basic laws called the first and second law of thermodynamics. The first law states that energy can be neither created nor destroyed, although its form can change, as in photosynthesis.

The second law states that whenever energy is actually used to do work, some of it is lost. This is not complicated. It is as simple as this: A tallow candle is a concentrated form of energy stored in the fat cells of the animal from which the candle was made. The energy stored in these cells comes from the green plant food the animal ate. The end of the cycle is reached when the candle is burned and the energy is dispersed in the form of visible light and heat. There is no way to reassemble what is left, to make either animal, wax or green plant again. This is what happens when resources such as coal, oil and gas are burned. They are *nonrenewable* resources.

Ecosystems do not function like energy. For the most part they completely recycle everything within them to create a constantly renewable, living, active community. Organisms at every tropic level (or feeding level) use energy to maintain their metabolism, to grow and to reproduce. According to the second law of thermodynamics, energy lost at each level becomes unavailable to the next level. (See Figure 12 on page 53.) More energy, therefore, is available to plants than to herbivores and so on up the pyramid. This explains, among other things, why there are many more organisms at the bottom levels of the pyramid. For instance, more prey than predators, more sparrows than hawks. The "Rule of 10" applies here just as in the food web, because this is the food web at work.

The second law of thermodynamics also explains the difference between energy flow and the nutrient cycle. Energy takes a one way trip up the food chain and eventually becomes "unavailable." Nutrients, in contrast, move in a cycle. For example, the grass is eaten by a rabbit, the rabbit eaten by a fox, the fox by a bobcat. When the bobcat dies, it is eaten by a vulture who may soon sit in a tree by the grass and whose droppings fertilize the grass. Nutrients do not travel in a one way flow, but in a circular pattern - the nutrient cycle. You will note, however, there is no energy cycle as such. Instead, it is most often referred to as the flow of energy within an ecosystem. Can we see this flow of energy within an ecosystem?

To answer this question, let's take a trip to one of my favorite Louisiana spots. See if you can recognize what we are actually watching in terms of the flow of energy. Some of you may actually live in such scenes every day of your lives.

Imprinting is perhaps the reason I choose to sit in the car to view the show, the windshield providing a familiar frame - like the shape of the TV or movie screen. I prefer to think it is the mosquitoes, players themselves, that determine my choice. What I see is beauty and life in motion.

From the direction of Barataria Pass off Grand Isle, the *Night Star* enters Bayou Rigaud. It is heading to the dock with its shrimp catch. Behind it, waving like a flag, a curtain of gulls are soaring, dipping, diving as the man sorting the catch tosses overboard the dead or undesirable. Death feeds life. Life moves on. The gulls are not the only takers. Moving like dark and rolling waves come the porpoises eager to collect their share of this giveaway food. While down below, many more mouths await what rains down.

Now an oyster boat comes up the bayou and slowly moves into the slip beside my car. The deck is covered with stacked gunny sacks of oysters. On the road behind me a noisy freight truck, obviously bound for city markets, moves in and begins to load sacked oysters via an equally noisy conveyer belt - another form of energy at work.

To the west the sun is beginning to slip quietly toward the horizon. A passing great white egret casts black shadow pictures on this huge bright orange tail light of the departing day.

Toward the east and the pass, Nature's own fishermen, the droll-faced brown pelicans are busily plying their trade. Their awkward grace touches me and I watch in wonder their spectacular dives. I am moved, too, by the long, low graceful line in which they glide silently over the water. It seems fitting this vanishing species will make its attempt to comeback here. They are home. I wish them well.

Just beyond the dock in front of me, mullet jump and splash, and shine like sparks of living energy. On the dock, two Louisiana heron and a boat-tailed grackle are having a loud and raucous quarrel over two crusts of bread. I am amused by the gawky stance, sinuous necks, glaring eyes, and windblown feathers of the heron. Indeed , it is count and counterpoint to the swift and more agile grackle. Finally, a swift move by the grackle, followed by a quick thrust of one heron's beak and the bread is gone. The loser departs the scene with an angry squawk. His dagger-like bill seems to stress his apparent mood. Life is so alive. And while energy is the chain that binds, even it is not the whole picture.

In the last rays of the sun, two raccoons on the opposite shore of the pass go busily about gleaning their meal along the water's edge. A few feet offshore, "bec á ciseaux," the black skimmer, cuts the water with the lower part of his scissor bill. The bird, like a miniature fighter plane, strafes his target. The wake of his bill is defined by a V-shaped streak of light on the quickly darkening water. Night brings its quiet to the bayou; yet, unseen and unheard, energy continues on its never-ending, whirling flow through our world.

Everything described here reveals visible examples of the role of energy in an ecosystem. There's more than meets the eye here. What is happening beyond our vision, underwater and at the microscopic level, provides the foundation for every ecosystem:

1. Inorganic substances are non-living substances, such as dead plants and animal remains;

2. Organic compounds containing carbon, like carbon dioxide, and proteins and carbohydrates that link biotic (living), like algae, and abiotic (non-living), like detrius materials.

3. The air, water and substrate, or foundation environment, include the climate and all other physical factors. The water bottoms would be in this category.

4. Producers in an ecosystem are living organisms, largely green plants that use photosynthesis to convert energy, like phytoplankton of various types.

5.   Macroconsumers in an ecosystem are chiefly animals that ingest other organisms, such as bottom dwellers and worms.

6.   Microconsumers primarily have the job of decomposing. They include heterotrophic organisms, like bacteria and fungi, that obtain energy either by breaking down dead tissue or by absorbing dissolved organic matter exuded from the plants and organisms. This decomposition activity releases inorganic nutrients that are usable by producers and also provides food for macroconsumers.

Bacteria are the main microconsumers here. They shuttle the matter of the marsh though myriads of chemical pathways. Organic compounds, once insoluble minerals, become soluble. All are sent whirling around the intersecting cycles of carbon, nitrogen, sulfa, phosphorous and others -- again and again and again.

Estuary systems pump away under the sun at an amazing speed. An average molecule of nitrogen may make the complete circuit from the dead body of an organism to the living body of another organism in a few hours. Molecules of phosphorous may undergo a completed cycle of transformation on the average of every few minutes. The speed with which these nutrients move within this cycle, as part of a brew of bacteria, diatoms, plankton and fungi, is the key to the abundance of all estuaries.

Air, formerly an invisible, odorless, tasteless mixture of gases, surrounds the Earth. Air, the breath of life, is 78% nitrogen, 21% oxygen and varying amounts of water vapor, carbon dioxide, ammonia, helium and other rare gasses.   Of the 107 known elements, only about forty are needed by living systems to maintain life. Many of these are needed in only trace amounts. Oxygen, carbon and nitrogen constitute the largest part.

Energy is not the only product of photosynthesis, the vast majority of oxygen in the atmosphere is also a product of photosynthesis. Indeed the humble green plant may have completely changed the chemistry of our planet's atmosphere. One theory holds oxygen gradually accumulated in the atmosphere over the millennium because more was produced through photosynthesis than was needed in oxygen-consuming processes. A more recent theory proposes some oxygen was stored in rocks or meteorites and released upon impact or after gradual disintegration. Regardless of how it happened, as the atmosphere became oxygen rich, Earth became more suitable for occupation by more organisms. By the same token, available oxygen built up an ozone shield that blocked out harmful ultraviolet rays, making the atmosphere suitable for even more varied forms of life.

While all ecosystems play a role in atmospheric control, wetlands, like rainforests, are lead players. In wetlands, decomposition proceeds rapidly and many important nutrients, sulfates and organic matter become gasses. The connection between solids, liquids and gasses is close here. There is no break in the cycle. Wetlands recycle solids and liquids into the atmosphere as methane, carbon dioxide, ammonia, and gaseous nitrogen compounds. Just as water cycles from the clouds, to the air, to the sea, the snow, the dew and back to the clouds, other vital elements of life also follow cycles. **Figure 19** shows diagrams of the most important one - oxygen.

Nitrogen is some four times as plentiful in the atmosphere as oxygen, but its cycle is more complex because relatively few organism can utilize it directly. **(See Figure 20, Nitrogen Cycle.)** Nitrogen is a necessary part of proteins. It is essential to animals as well as plants and is a major storage pool in ecosystems.   Blue green algae and other bacteria "fix" nitrogen- this means they convert the atmospheric form to a somewhat more complex inorganic molecule that can be used by plants. Lightning, photochemical reaction, modern fertilizer, plus nitrogen fixers transform it into forms usable by living organisms.

OXYGEN CYCLE

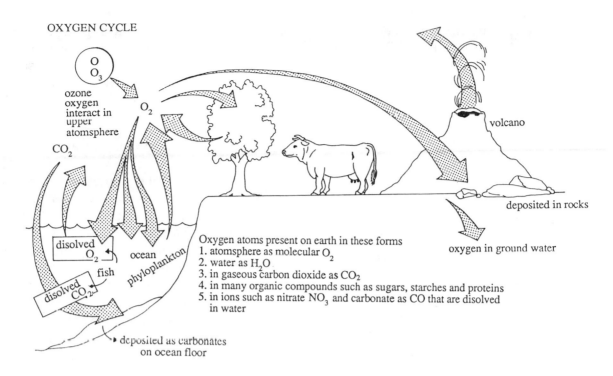

Oxygen atoms present on earth in these forms
1. atomsphere as molecular O_2
2. water as H_2O
3. in gaseous carbon dioxide as CO_2
4. in many organic compounds such as sugars, starches and proteins
5. in ions such as nitrate NO_3 and carbonate as CO that are disolved in water

Figure 19

NITROGEN CYCLE

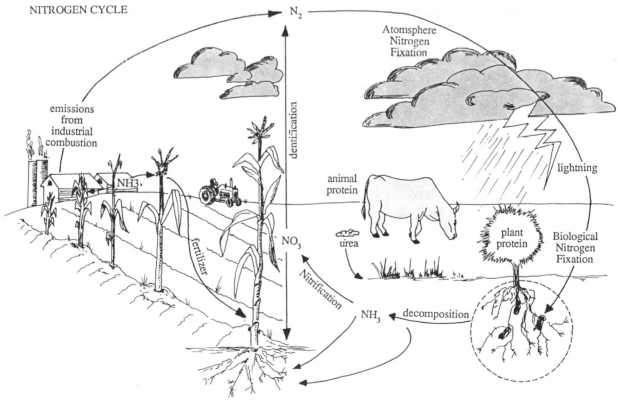

Figure 20

There is also a mineral cycle. Cycling minerals, such as phosphorous, calcium, sodium, potassium, magnesium or iron is much more fragile a process than cycling carbon, oxygen or nitrogen. In a healthy ecosystem, minerals and nutrients in the soil are absorbed by plants and carried to leaf tissue. The nutrients are generally assimilated, but most of the water is lost through transpiration. Most of the rainwater, therefore, must ultimately flow into stream beds. Since all ground water carries nutrients out of the system, a large increase in stream flow disrupts the mineral balance. **(See Figure 21- Phosphorous Cycle.)**

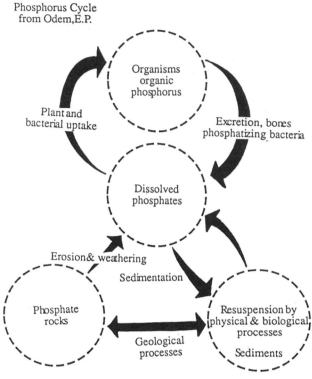

Figure 21

This information on the oxygen, carbon, nitrogen and mineral cycles can now be used to explain some facts about pollution. Many chemical discharges into water are diluted and seem to be assimilated, but in many cases the opposite occurs. What is known as magnification occurs. Pollutants an organism cannot metabolize or excrete, accumulate and are magnified at each level of the food pyramid. Each successive organism eats many times its own weight in a lifetime to compensate for energy used and lost; pollution, therefore, accumulates and is concentrated ten-fold. **(See Figure 22.)** This is frightening because so many people are unaware of this and feel that once pollution is out of sight and away from media attention, that is the end of it. But really, in most cases, this is only the beginning. Out of sight, pollutants, no matter how much they are "cleaned," will continue to create a path of destruction for many years to come. This no one mentions.

The primary source of energy, of course, is the sun and much of this energy is eventually radiated back into space. A long-standing theory holds the average annual temperature of the Earth remains relatively constant because the inflow and outflow of energy are equal. What is known as a "steady state" exists, according to one group of scientists. Recently, however, a new theory appeared which holds that the Earth is warming up, like a greenhouse. This theory is called the Greenhouse effect and holds that the Earth's temperature is increasing as the amount of carbon dioxide increases. Our Earth is not large enough to handle carbon dioxide (CO_2) in the volume we are producing it without upsetting the fundamental balance of the biosphere.

The sun also causes air to move by heating it. Heating causes air to expand and become less dense. Less dense air rises like a huge invisible balloon, the cooler and denser air rushing in to take its place. In this way, the wind is born.

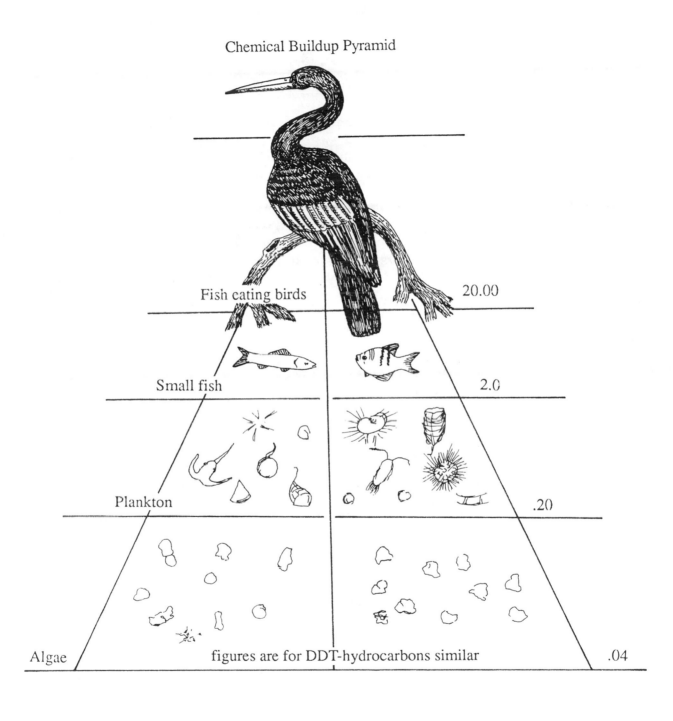

Chemical Buildup Pyramid

Fish eating birds — 20.00

Small fish — 2.0

Plankton — .20

Algae — figures are for DDT-hydrocarbons similar — .04

Figure 22

A View from the Heart

The Navajo Indians believed the lines on our finger tips were "tracks of the wind" left there by the wind when it made their ancestors.

When the Earth was born, so was the wind; yet of all natural phenomena, the wind is the least controllable. We have long realized its tremendous effect on our planet, from the shape of the eagles, to the shape of our coasts. All have been designed by the wind. Wind, constantly redesigns land, water, coasts, rocks, sky and clouds, and, if not man himself, then man's designs. Pyramid, tepee, race car, power lines, space age planes, all have been designed to accommodate the wind.

The wild ones still read the wind. Migration, although exactly how still eludes us, is in some measure effected by the wind. Old timers on either side of our flyways know that ducks sense a change in the wind. In turn, these watchers note the change in the ducks. One day they will say: "In the morning they will all be gone, every one."

Over the millennium, we have at times been blessed, and at other times, ill-used by the wind. We have had some success in learning to read the wind. Hunters know game travels with the wind and so keep it in their faces as they stalk their prey. Fishermen and farmers live and breathe to its rhythms, reading its changes, knowing how it will effect their work and their lives.

Yet today, except at certain times, many people have little to do with the wind. They get their weather from TV. They do not even think about wind, except when it outrages us in hurricanes or tornados. Yet it rules us still, just as it always has, and perhaps always will. How sad so few can still read the wind. Yet some still seek it, along the shore, in the high mountains, soaring on a glider, sailing a boat, or merely in a garden or murmuring grove. We may not all be able to read it, but we can still listen- for it speaks to us... What does it say? What does it do? It tickles the willows and makes them laugh. It sings gently to the sleeping calf. Yet it raises its voice over the thunder's roar and drives the wild waves to the Gulf shore. It cradles the newborn storm and helps it grow. Sometimes, it capriciously brings us snow. Timeless as always, wild as forever, blow the constant winds of change. Is the wind that blows on Mars, and other distant worlds among the stars, a silent wind because there is no one there to listen?

Listen...cypress trees are beautiful singers when the wind is near.

Bayou Country Ecology

Nature's ways are not simple, but they are thought-provoking.

In working with computers, various phenomena of ordering given to computers show patterns that keep creeping up in events as unlikely as the formations of a thunder storm and the flow of blood through an artery. Even something as seemingly random as the curling of cigarette smoke into nothingness, turns out to follow a predictable pattern. Years ago, it was believed that all happenings, whether of will or of Nature, were determined. They were *determined* by preceding events. But for those who lived in such seemingly unstable places as barrier islands, *Determinism*, as it was called, was not acceptable. They felt instead a dynamic energy at work that went beyond the seemingly random energy of wind and wave. Although some knew of Brunn's Rule (see page 23), others felt there was no order; that their right foot was saying order while their left foot said chaos. They felt strongly these reactions were more than the results of previous events. In the case of barrier islands, what appears to be unstable or chaotic may be order we are not yet able to perceive.

These feelings gained support when, in 1934, Werner Heisenburg proposed what is now known as the Uncertainty Principle. This principle states there is true chaos in this world; it even has an orderly mathematical prescription. Weather forecasters felt vindicated! It had long been believed that as soon as computers and various other devices were perfected, weather forecasters would be able to always accurately predict the weather. Not so.

Heisenburg's theory proved the stability of large scale ecosystems depends upon the existence of internal chaotic instability. We now know chaos is spread around the world, making it impossible even to accurately predict changes in Nature. Such chaos is the counterpart of order.

The remarkable thing about healthy ecosystems is their apparent ability to absorb such destabilizing influences. In fact, their stability in some way depends upon the existence of this chaos.

Not too many of us are prepared for something as thought-provoking as this. But equally thought-provoking are the questions that arise as we face decisions to try to save Louisiana's wetlands. These include:

Q. What will happen if the loss of biodiversity continues to increase?

A. As biodiversity decreases, there will be more herring gulls, fewer ducks; more weeds, fewer wild flowers; fewer butterflies, more roaches. Essential parts of the genetic library we hold in trust, this vast storehouse of organisms, millions upon millions of species and distinct populations that constitute our support system, will become more and more difficult to preserve. But the beginning of the age of genetic engineering is upon us, and while some are frightened by its implications, now more than ever our goal need be not "no net loss of wetlands," but no avoidable loss of any genetic material to assure scientists as complete a library of genetic material to work with as possible. Even now, attempts to clone a new and more stress resistant species of grass could possibly do a great deal to save our wetlands.

Q. How, exactly, is this downward spiraling causing the problem?

A. In self-regulating systems, each change, or turn in the normal spiral, results in an increase in the competition for adaptation and an increase in opportunities for new adaptations. As the opportunity for new life forms increases, the number of niches and the creatures to occupy them increases causing all species to compete in new conditions. But once the spiral is reversed by Nature or by man, this highly interrelated complex of species is more susceptible to rapid alteration of conditions. Each species lost then threatens others that depend on it.

A View from the Heart

Q. What is natural and what is right?

A. This may well be one of the hardest to answer. We need to be wary of using a single species or group as indicator, for several reasons. For example, when fertilizer washed into Chesapeake Bay it killed much of the grass that grew in the bay. This grass was prime food for ducks. The duck population went down. But, these same fields from which the fertilizer had washed into the bay produced a rich area of corn fields. Geese that winter or live in the area feed in these fields. The geese population went up. What is natural and what is right in a case like this? Yet what this does show is the resilience of natural ecosystems. Often the biomass appears to remain the same regardless and there can be great difficulty determining if a given area is actually coming or going.

One way to make this kind of determination is to establish a base line beyond which we dare not go. A base line is a pre-set standard against which to measure progress or decline. A base line holds us accountable, pushes us to also establish a bottom line, a level of degradation below which the situation is simply unacceptable. Unless this is done, each new generation will accept a lower bottom line, edging nearer and nearer the ultimate bottom line - extinction of a species or ecosystem.

Protected natural areas, off limits to mankind, may be essential tools in understanding and maintaining diversity as well as health and continuation of ecosystems.

It is not going to be easy to change how people here in Louisiana view their environment. People feel very strongly that the environment belongs to them and that it should be made to work for them. Their fathers and their father's fathers fought hard to win this land. Like most people who work close to the land, they find it difficult to listen to someone telling them what to do with something they feel is already theirs. They hold onto a belief that what their forefathers did was correct. It fit the times. It brought to them, the present generation, strength and determination. But, the future of this bayou country is up to these truly caring people who love their piece of the world. The traditional way of doing things has to be adjusted through education. Bayou people have a long history of overcoming difficult odds. They know compromise and accommodation. Louisiana bayou people truly do have a passion for life and for their area. When the time is right, they will do what is necessary to preserve this environment they love. They will be willing, I believe, to listen and heed the wisdom of Jonas Salk: "What is more important than survival of mankind is that we keep evolving as a species. If we are to survive as a species we must learn from our mistakes, as nature does."

No matter where on earth we are, there is a long journey ahead of us. If we are ever to have a planet ecologically at peace with itself, a journey towards understanding must begin inside each of us. There is no other way. The more we learn, the more we will find there is to know. Life is a circle, a very precious, relatively small circle in which we are but a small, single living part. From earliest times, humans have felt its magic. From such concrete evidence as Stonehenge, to the Sacred Loop of Nations of the Indians, we find man has long known that what Nature does, is done in a circle.

Omar Kiam, Persian astronomer and poet of the eleventh century, sensed this: "I come like water, and like wind I go."

90

Bayou Country Ecology

We are lucky that so many brilliant minds of the past and present have left us their words as a guideposts along our journey. Some hundred years ago, Aldo Leopold wrote:

"...When we see land as a community to which we belong
we may begin to use it with love and respect."
– Aldo Leopold

The Earth is approximately 8,000 miles in diameter. From its crust (the lithosphere) to the outer reaches of the ionosphere, which includes the troposphere and the stratosphere, it is several hundred miles.

Sounds tremendous, doesn't it?

Think again.

The biosphere, which is the only part of the planet that man can explore without some type of life support system, is actually an area barely five miles in width. **(See Figure 23.)** This area supports not only the roughly 5,321,000,000 people who presently populate our Earth, but all the creatures, plants and water that sustain them. Because 70% of the Earth is water-covered, then only 30% can actually be inhabited by man and land-dwelling animals. How precious life is...and how small the space on Earth that supports it.

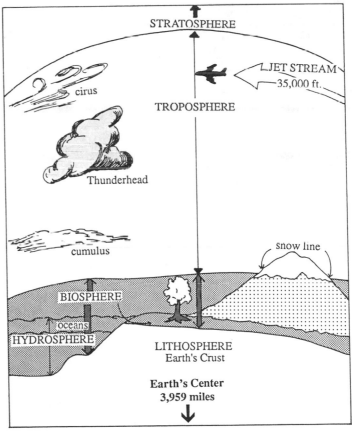

Figure 23.

Bayou Country Ecology

9

The Oil Legacy

*"We are hollow men...
Our dried voices,
when we whispered together are quiet
and meaningless
As wind dried grass."*
- T. S. Eliot

Millions upon millions of years ago nature produced oil under the surface of this land and coastal shelf. In 1893, Huey P. Long was born within its borders. Because of these two facts, Louisiana will always be different than it would have been if neither of these events had occurred.

Louisiana and its people have a natural love of politics, perhaps stronger than any other state, possibly fostered and encouraged by the Longs and the political empire they built. In 1934, Huey P. Long, for reasons unknown to the public, followed the old adage - "If you can't lick 'em, join 'em." He and several friends formed a company that bought state mineral leases and re-sold them for large profits to out-of-state oil companies - Texaco, Gulf, and others. Oil had been discovered here earlier - the first gusher came in 1901 - but after 1934, Louisiana was literally one state-wide "Boom Town." With Long and his partners giving and accepting the rings, the oil companies and Louisiana politicians, despite frequent squabbles, had a happy marriage for fifty years - a golden marriage indeed.

On their side the oil companies and related businesses have been good providers, being among both the largest employers and the largest taxpayers. They shared the profits well and won practically the full support of public and politicians alike. The building of the first offshore oil rigs in the Gulf in the 40's issued in yet another wave of prosperity. Following on the heels of oil development was another giant - the chemical industry. It is only recently that this healthy and happy state began to have money problems - a shrinking tax base. Oil companies began shipping more offshore oil through pipelines that bypassed the state. Problems continue to mount as more and more signs indicate the oil fields are beginning to play out. The interest of oil companies seems to be fading fast as the once mad rush for the state's oil begins to slack off. Louisiana is beginning to see that much of her natural ancestral dowry has flowed out of state. In the span of one man's life, Louisiana has drained the bulk of its known

93

oil. The end won't come tomorrow or the next day, or the next after that, but come it will. It is as inevitable as the passage of time itself; already many operations are pulling up and moving to federally held territories deep in the Gulf of Mexico.

What has been a marriage of convenience from the start is not likely to fall apart easily. Each side will wait till the last good-bye is said.

Most adults now living in coastal Louisiana grew up in the shadow of the oil companies. For many, it is the only life they have ever known. Like people anywhere, they don't want to give up. People who have worked in mining or steel all their lives and who know the faults and good points of their company as well as they know their own, can surely identify with the people of coastal Louisiana. Both have grown to accept the circumstances of their working lives and will not voluntarily have them changed. Here there is another point to consider. Instead of being affected by depression in only one industry, Louisiana is impacted as well by the sharp decline in the many and varied types of companies and businesses oil and gas has spawned.

Neither side can be blamed for the present situation. Times have changed and that is what we must come to terms with. We need to take good care of what we still do have and make the most of our remaining dowry from Mother Nature that has not been spent and is restorable. There are also excellent prospects for the future of gas reserves. Whoever would say we can't dry our eyes on the torn lace of our wetlands, doesn't know Louisiana people very well. Already people are beginning to re-assess the future and to face it with renewed vigor. There remain hard choices to be made, hard questions to be asked and answered. Much needs to be learned by all. What exactly is Louisiana's environmental legacy from her "black gold "days? In what condition is the remaining "green gold "of our marshes?

The Environmental Legacy
of Oil and Gas Production
Loss of Marsh Land

There is a crisis in Louisiana's marshes: land loss is accelerating. What part of this loss can be connected with oil industry activity? Canals. Few people, even those in Louisiana, realize just how many thousands of oil field canals crisscross the marsh lands. There are an estimated twelve thousand miles of these canals. They adversely effect the marsh by interfering with sheet-water flow, allowing destruction by wave action, and reducing nutrient exchange. Boat traffic throws a wake that eats away fragile vegetative layers. But by far the most serious damage has and continues to be caused by salt water intrusion. We already know the effect of salt water on marsh wildlife. The extent to which canals contribute to the resulting marsh loss is also enormous.

Digging these canals created soil banks, deposits of material dredged to make the canals. Naturally occurring elevated coastal areas, such as natural levees and cheniere ridges, provide a stable area for many high ground plant species and increases habitat diversity. But man-made banks imposed on the system severely impair water movement. Before these banks erode and subside they do permanent damage to adjacent areas of land which sink and then fill, converting to bodies of water. This finding should provide a final answer to questions about whether canals are a help or a hindrance to marsh stability. Yet questions on the impact of man-made canals and banks continue to be asked. Some Louisiana officials want to use dredged material to rebuild the state's coastal marshes. So once more the meetings begin.

Such problems are not indigenous to this state, of course. Nationwide, there is a multifaceted collection of overlapping problems connected with environmental protection, economic growth and management of coastal resources and activities.

In some ways the winding down of the oil boom in Louisiana may be a blessing in disguise in that without the dredging of additional canals and channels, Nature, over time, may be able to re-establish equilibrium. She has a way of often doing that, given half a chance. This blessing would not, of course, eliminate the dumping of chemical wastes that have accumulated over the years from oil industry, chemical plants, and the disposal of hazardous wastes in Louisiana.

Bayou Country Ecology

Land Subsidence

Subsidence can, to some extent, be the result of natural sinking or settling of earth over time. Such movement provides grist for academic studies. None of these can account for the subsidence acceleration in recent years. Some man-made causes of subsidence include compression caused by the weight of objects such as buildings, pile structures, artificial levees or land fills. The extraction of oil, gas, sulphur and water from salt domes is also a contributing factor. Former studies indicated sinking to be averaging eight inches a century, which was more than compensated for by silting. Today, subsidence is estimated at a rate of two feet per century.

Aging Oil Fields

Many old and useless wells dot the Louisiana landscape. The draining of oil and gas from reservoirs over the years has become a problem. Aging oil fields produce large volumes of salt water, an average of eight times as much salt water as oil and gas. This water from deep within the earth is often seven times as salty as the ocean and it can contain many potentially toxic chemicals.

Routinely for years, some of this brine was dumped illegally. Today it is hoped that most is injected into disposal wells of which Louisiana has more than 4500. The state did not begin to monitor these disposal wells at all until 1982 and then it was not happy with what it found. Open pits continue to be used and are not closely regulated by the state. Thousands of such pits are now filled with brine and a soup of chemicals. Waste sludge also is pumped into abandoned wells, sometimes spreading over adjacent land.

Once again, such things are not indigenous to Louisiana alone. The disposal of hazardous waste is a national problem and a national disgrace. As of 1990 regulations forcing oil and related industries to stop discharging polluted waste water into coastal marshes and bays has been approved by the Louisiana legislature. These rules will be phased in over a period of four years.

Oil companies object, of course, claiming it will cost them nearly $1 billion to drill wells to be used to inject waste water into the ground. They further state production will decline and the state will lose $200 million in severance taxes. This same argument was used in 1985, when regulations were passed forcing oil companies to dig the original 4500. Once again, there will be endless studies by both sides, studies with findings that amazingly always favor the initiating party. What are the people to believe?

Remember, the time frame for environmental damage is long. For example, once any of these chemicals reaches the aquifers - the permeable layers of material through which ground water moves and which supply about 85% of public water - nothing can be done. This could well mean that generations to come will be exposed to carcinogens, PCB's, heavy metals, hydrocarbons and other toxins that cannot be eliminated.

Many reports are done on such things and for one reason or another, in one way or another, never become public knowledge. This, too, is a country-wide problem. Perhaps this could be called the "Politics of Confusion." No one can blame the average person for being confused when one set of "authorities" comes out in favor of some plan and another set of "authorities" comes out against it. Profit is often the deciding factor. The classic example in Louisiana is the controversy over shell dredging in Lake Ponchatrain. This profitable business was not good for the ecology of the lake, yet it was allowed to continue for years. The bottom line is that now the Lake remains in serious trouble. The "Politics of Confusion" succeeds through manipulation and ignorance. The people who stand to lose the most

A View from the Heart

96

are the last to know and to understand. The public needs to become aware. Above all else, we need to teach our children to think - to really think - for they will be facing problems more complex then any we have yet encountered.

Hazardous Waste

Hazardous waste is a part of Louisiana's oil legacy. The issues surrounding hazardous waste problems have resulted in introducing to the public a collection of highly technical words and concepts that can be confusing. When we hear of a new oil spill, often it is described in terms and words that, for many, have little or no meaning other than the familiarity that results from constant repetition. Words like carbons, hydrocarbons, organic compounds, and dispersants. What do they mean?

The world of chemistry is a vast and complicated one. But by defining a few terms we can help ourselves understand something of what goes on.

Everything in the world is made up of basic elements. These elements combine to form the tissues of all living things, the substance that makes up the earth, and even the air we breath. All material on earth is either an element or a compound of elements or a mixture of compounds and elements.

A compound is composed of two or more elements always in the same definite proportion. Pure water is an example: it is always composed of two gasses - hydrogen and oxygen. Sounds simple, but indeed it is far from it. Just consider that hydrogen becomes an explosive gas when mixed with oxygen (another gas) in the air. Oxygen is a gas that is needed before anything can burn. Yet H_2O is a liquid that will not burn and is indeed used to put out fires. Compounds, therefore, have properties totally unlike the elements from which they are made.

Most of the compounds in the world are organic, which means they contain carbon. This large number of carbon compounds is due to the fact that carbon atoms are able to attach themselves to each other as well as to atoms of other elements. Hydrocarbons, compounds containing only hydrogen and carbon, occur in nature as natural gas, petroleum, turpentine, and asphalt. Industry creates hydrocarbons such as gasoline and kerosene. Acetylene, a simple hydrocarbon that burns with oxygen, is used in welding. Methane, a by-product of natural gas and coal gas, is a hydrocarbon that can cause mine explosions.

All of the compounds that make up oil (petroleum) are hydrocarbons. Understanding a little about them will help us better understand what happens when an oil spill occurs.

Since oil is less dense than water, it tends to float on water and form a slick. A common assumption a decade ago was that after a spill the slick "weathered," evaporated, decomposed and degraded, becoming less toxic. These processes do make it disappear from the human eye, but research has shown that it not only remains toxic, it becomes more toxic as it enters the food web.

Hydrocarbons in petroleum rarely dissolve. Those that tend to dissolve the most also tend to be the most toxic. When these highly toxic hydrocarbons are dissolved, they then become more readily available for uptake by organisms.

There are many ways hydrocarbons reach water bottoms. Oil can be mixed with water in suspended droplets and weather conditions can drive it deep into the water and onto the bottom. Oil and sediment can combine, and being heavier than water, sink to the bottom. A third way is in the fecal pellets of certain species of zooplankton known to feed on hydrocarbons. Still another way is through chronic low level discharges of oil from certain industrial plants. Some of these new substances are potentially toxic.

Research has shown that doses of toxic material with concentrations as low as 10% can inhibit the growth of various types of phytoplankton and reduce reproduction in some marine animals. It can also interfere with the breeding behavior and communication between animals. While adult fish are fairly tolerant of oil, fish eggs and young larva are easily poisoned by even low concentrations. Shell fish at any age or stage are particularly susceptible to oil contamination of any kind. On the other hand, certain opportunistic species are hardly bothered by the presence of oil, while others are affected differently, depending on the species, the degree of tolerance and the stage of development. Concentrations of 20% oil in water result in subtle changes in complete ecosystems beginning with a change in phytoplankton. This affects the type of bottom dwellers that eat the plankton, the fish that feed on the bottom dwellers and ultimately the human community. This is not an overstatement. Knowing this, when do you think the damage done in Prince William Sound, Alaska and the Persian Gulf will truly be "over"?

Remember about giving Nature a niche? Well meet Capetella capitata, a little, blood-red earthworm able to survive in oil contaminated sediment. According to a Texas A & M University ecologist, some 40 colonies of clams, mussels and brightly colored tube worms have been discovered living around oil seeps in the Gulf of Mexico. As deep as a mile below the surface, far beyond the reach of visible light, they feed on bacteria that are able to convert hydrocarbons to energy. Among these creatures are many species new to science. Spooky isn't it, what Nature can do?

Drilling Mud

Do you know what drilling mud is? During the drilling process at a new well, tons of viscous material called drilling mud is injected into the hole. Such mud may contain as many as 500 different chemicals that act variously as foamers, de-foamers, flocculents, thinners, and emulsifiers. Some of the chemicals in drilling mud are asbestos, formaldehyde, aluminum, carbolic acid, caustic soda, chromium, barium, titanium, and arsenic. Lead and iron are sometimes added to increase weight. Lubricants, such as asphalt laced with phenol, are often added to reduce drill-bit friction. There's more. For years after the original drilling, regular doses of such things as ammonium bisulfite and zinc chromate (corrosion inhibitors) and acid compounds, plus more formaldehyde (an anti-bacterial agent) are added. What comes up is often harsher than what went down, as it will pick up naturally-occurring chemicals, such as mercury, radioactive isotopes of potash, cobalt and others.

Guess what? Louisiana doesn't classify drilling waste as hazardous.

The manufacturing of drilling mud is big business. Louisiana's petro-chemical industry spends over one billion dollars a year on it. In Coastal Louisiana, where, due to deep pocket formations, wells tend to be very deep. Drilling a typical 10,000-foot well may require a million pounds of mud. Of this, as much as one third remains underground. The rest is supposed to be recycled or disposed of. Dry holes are commonly used for disposing of un-recycled drilling mud since hauling it away is an expensive process. No one will ever be able to determine just how much has been discarded in this way improperly. But, is this approach any worse than the method most often used, that of spreading it over near-by land, a time-honored method of disposal in South Louisiana?

If that seems acceptable, read what that wise prophet Rachel Carson, had to say in **Silent Spring** many years ago. Soil contamination remains for years and years. Louisiana delta farmers are today using contaminated land to grow food.

Controversial as it is, the Environmental Protection Agency did propose to regulate drilling waste disposal. When the oil industry lobbyists cited the staggering costs of such regulations, Congress, in 1980, exempted oil and gas from regulation. A study was ordered, but the EPA never pursued it. California took a different view of the situation and requires that all drilling compounds with certain levels of metal or toxic materials be disposed of just as is any hazardous waste.

There is no question, though, that the potential hazard of oil field waste has been glossed over in Louisiana. What research that has been done is anything but reassuring. All kinds of boats use diesel fuel and it also is one of the leading drilling lubricants. But, a few years ago a published paper did show that diesel fuel, even in tiny quantities, can kill mysid shrimp, a tiny brine shrimp occupying about the same chain of the food web as the copeopod.

If there is a connection between all these things and Louisiana's high cancer rate, no conclusive studies have been done to prove it. The state's chemical industry, instead, did a study a few years ago of the effect of life style on health. In that study, smoking, drinking, and low dietary fiber were blamed for the region's high cancer levels. Meanwhile questionable drinking water and high cancer rates remain a fact of life in South Louisiana.

This is Louisiana's legacy from the petrochemical industry. Time and tide wait for no one. What are our options? Where do we go from here? That's easy - ahead. The "green gold" of our wetlands is still about us, a little worse for wear, no doubt, but nothing that love, learning, and hard work couldn't remedy - in most situations. Our "black gold," as we well know, is not renewable, but our "green gold" is, and, with luck, it will remain so for many, many generations to come.

Bayou Country Ecology

"Far in this ethereal sea
lie the Hesprian Isles
Unseen by day
But when darkness comes
Their fires are seen from this shore
As Columbus saw the fires of San Salvador."
- Thoreau

Out there beyond the horizon lies another world. A world as foreign to Nature's world as if it were from another planet. Money built it. It operates for oil money. Money keeps it running. Money is why its workers stay. Money is God out there. But chance is still the name of the game, in this world of the offshore worker.

People hear the work on rigs is long, hard, rough and dangerous. They say it breeds a special camaraderie and few talkers. You gather facts where you can.

Somehow you think of anything at sea as being quiet. Yet here is a world of noise, of metal against metal, of man against man, shouting and being shouted at. A world of dirt, mud, paint, cement, grease; of being too cold, or too hot, or too wet. It is blowing wind, and pitching and rolling waves when it is stormy. Not so much frightening as making the work harder and more dangerous. Long hours mean having trouble keeping eyes open and trouble keeping them shut when you finally reach the bunk. It means good food and good pay but bad times. It means climbing down inside rig legs to inspect the ballast 50 feet below the sea. It means hanging by a safety belt 100 feet over water, feeling maybe something like a spider on a steel thread. Coffee, cups and cups of it, and, in some cases, drugs, keep men going, unloading boats, bundles of pipes, power tools, drilling mud, drill bits, and mail. Offshore oil work means having danger for a constant companion.

But men are not the only creatures who find danger on the rigs. Offshore, money attracts and holds men, but light - man made light - is what attracts and holds others creatures and often draws them to their deaths. Perhaps it is only a swarm of dragon flies or migrating butterflies, but more often it is migratory birds that are the victims. At times such creatures may fall like dead rain on the deck. A broom sweeps them into the sea and only Nature mourns her loss. Perhaps tonight some lovely white egrets circle a rig, again and again, hour after hour, held captive by the lights. Only morning releases the birds.

A great silver bird appears out of the blue, hovers, then lands on the deck. Men pile in. Their bird takes off. Now they too are free...but only for seven days.

10

Realm of the Riddle

"Nothing is simple and alone.
We are not separate and alone.
The breathing mountain, the living shore,
each blade of grass, the clouds, the rain, each star, the beasts, the birds,
and invisible spirits of the air - we are all one indivisible.
Nothing that any of us does but affects us all."
– Frank Water (The Man Who Killed the Deer)

For many years, most people believed that "charity" began at home. You took care of your business and the other fellow took care of his. This was especially true of people who were fiercely independent. It was good and it fit the times. But a funny thing happened on the way to now, the meaning of the word "home" changed. We have only recently come to the full realization that what happens half a world away can effect us in just a couple of days in ways that we had never dreamed of before.

Life grows more complicated every day. We are faced with decisions - everyone of us, serious decisions, personal decisions, no longer simple decisions like fishing and hunting regulations. We can see now that we no longer have the luxury of thinking that all that matters to us is what happens in our own area. Life has become a riddle.

What is a riddle? It is a puzzling question that exercises our ingenuity to discover its meaning. Up to this point in our exploring, all we needed was an occasional flight of the imagination. What we need now is a "flying carpet," for we will be journeying literally to the ends of the earth and back. We will be journeying to the "Realm of the Riddle." The question we seek the answer to is — Must we care about the world's problems as well as our own?

Here's your ticket for the carpet ride. We're off!

Earth's biodiversity, its billions of genetically distinct populations, whether on land or sea, is the "capital" on which all of Earth's creatures live. It doesn't take a financial wizard to know it's wise to live on the interest and retain the capital, whether in the world of finance or ecology. Diversity is our only hedge against the future.

A View from the Heart

Along the way here's something worth thinking about: Do global weather disturbances, overpopulation, and natural resource depletion, which includes the ozone layer, tell us something we cannot deny? Don't such symptoms suggest that the life-creating ecosystems of the world are in serious trouble? Such warning signs from our Earth does not mean the world is going to end tomorrow, but they do warn us the Earth is in trouble and we had better start doing something or we are going to be in trouble, too. We have bent Nature's laws as far as they will go. We no longer have the luxury of ignoring the world's problems Perhaps we never did - perhaps that has been the problem all along. We have failed to realize just how much "things come attached." We are learning the bitter truth of this - as is all of the world. To a great extent we can most affect our own area, but we need to be aware of what is going on elsewhere. We have arrived at the first part of our ecological riddle...

Destruction of the Rainforests

Due south of Louisiana is part of the Earth's most highly diversified and valuable natural resource - a band of tropical forest that circles the globe from Africa to Asia, to the Americas. There are two types: moist evergreen forests and moist deciduous (shedding leaves annually) forests. Throughout history these forests have been sources of food, fuel, medicines and countless other products. They support mankind and his environment by protecting soils and crucial water supplies. Their diversity is unbelievable; for example, 700 species of trees have been identified in one 25-acre tract of a Borneo rainforest - as many as are found in all of North America, according to **Conservation International**. It is not surprising to find that in these forests, which occupy only 6% of the Earth's surface, are 60% or more of the Earth's species. This means if this habitat is not protected, a quarter of all known life on Earth could be pushed into extinction within our lifetime.

If the Earth has a supermarket, this is it. Unfortunately, we have hardly begun to truly inventory its treasures, many of which may be extinct before we even learn they exist. Scientists estimate we are losing about two species a day. How? Quoting again from **Conservation International**: "In order to meet their daily needs the people of tropical developing nations often have no choice but to overuse the natural resources on which their long term survival depends. They simply do not have an alternate source of livelihood." Sound familiar? But there is a big difference. Unlike us, they truly live in a world of complete poverty, overpopulation and constant hunger.

Aside from purely humanitarian concern, what does this mean to those who do not live in these areas? For one thing, rainforests control the Earth's climates by acting as natural sponges that absorb much of the CO_2, the main cause of the Greenhouse Effect. Tropical forests also help stop massive climate changes because of their effect on global water cycles and because of atmospheric circulation patterns.

Another thing to consider - only a small percentage of plants in rainforests have ever been screened for chemicals that could be useful as medicines. "So what?" you might ask. Well .. a full one-third of all modern prescription drugs contain compounds originally found in plants. These include quinine, digitalis, and a leukemia and diabetes fighting drug derived from the Madagascar periwinkle - cousin to the periwinkles that grow in so many Bayou gardens. No one knows what wonders might still be out there. And that isn't all.

Bayou Country Ecology

The world's most basic staple - grain - depends on fewer than two dozen species (corn, barley, wheat, etc.) We need to search for new sources of potential food crops to feed the ever-growing population of the world. One such source is already known. Plants of the amaranth family, the same family as the garden variety cockscomb, can produce an edible grain. The problem with the amaranth is the breeding stock might be destroyed as diversity in the rainforests disappears. Since we know that all organisms are functionally part of an ecosystem and that the ecosystems need continued diversity, we can see why diversity in rainforests is indeed Nature's grand tactic for survival there and everywhere.

And still that is not the end. There is another part to the plant story that connects with this. The "homeland "of each food crop is of critical importance to plant breeders because it is where the greatest genetic diversity of each species can be found. Diversity is, in effect, a safety net that insures pests, diseases, or environmental changes will not plunge presently -used species of plants toward extinction. By protecting these homelands, it is possible to locate and re-introduce original stock, to work toward developing new versions of presently -used species, and to serve as a backup for endangered species. Corn is a good example. Today entire nations are becoming dependent on genetically identical corn plants. True, they are uniformly identical, uniformly highly productive, and uniformly excellent. They are also uniformly vulnerable to any mutant disease or pest that zeros in on them. But, unlike some food crops, the wild ancestor of modern seed corn still exists. To protect tomorrow's food crops scientists need the secrets locked in ancient, or, as yet unused varieties, possibly available only in places like ...vanishing rainforests.

"...With every loss of species we threaten the diversity of life that both feeds on and nourishes other life in the convoluted ballet of existence. When that happens, life is, and we are, in the deepest and most resonant meaning of the term, impoverished." -T.H. Watkins

But, should we care about vanishing species?

The most recent report on the 405 endangered species in the United States today, according to the U.S. Fish & Wildlife Service, shows that 3% of all insects are endangered; 4% of reptiles; 9% of mammals; 12% of snails, clams and crustations; 13 % of fish; 14 % of the birds; and, most disturbing of all, 45% of the plants.

More riddles - Sea Pollution

We of Coastal Louisiana say we love the Gulf. We know about pollution, we see it on our beaches and in our water. Have we already accepted this pollution as just "the way things are?" Do we question it? Do we realize the waters of the Earth are very vulnerable? Here are a few facts.

The chemical soup of the sea, something like the "gumbo" of our marshes, is being constantly stirred in its huge kettle. New elements constantly enter from land via the rivers. So why doesn't the pot overflow? While it never boils, it acts like one that does and uses evaporation to enter into the water cycle. The total mass of ocean water has remained more or less unchanged and in a stable state for millions of years. However, although the water is removed through evaporation, the salts are not. Why, then, doesn't the soup get too salty? Since we know it remains in a steady state, obviously excess salt has to be removed, but how? One simple way is through ocean spray, thrown into the air, that

105

evaporates leaving behind tiny air borne particles of salt which may remain in the atmosphere for days or weeks before rain brings them to the ground. Any gardener living close to the beach can tell you where many of the particles end up. Salt spray is the bane of their existence.

Scientists do not feel the salinity of sea water has changed much over geological time, but has remained fairly constant. Ocean water is a chemical mixture of 96.5% water containing six salt chemicals Chloride(CL), Sodium(Na), Sulfate(SO_4), Magnesium(Mg), Calcium(Ca), and Potassium(K). Other chemicals, as many as 75 or more, compose the remaining 3.5%. Some of these chemicals are buried in the sediment of the ocean floor. Some are deposited in nodules on the ocean floor. For example, manganese forms nodules composed of iron manganese and other substances. Phosphates also accumulate in nodules, described as potato-like, which brings to mind a funny analogy. What cook hasn't heard that one way to get too much salt out of your soup is to add potatoes?

Oxygen (O_2) and carbon dioxide (CO_2) are gasses that move through the ocean waters and into the atmosphere. There is constant exchange of these gasses at the surface level of the sea.

Oxygen is used by deep water animals for respiration and by bacteria for their work. What is unused moves from the ocean floor into the atmosphere. Concentrations of oxygen at any given place result from a balance between the processes that produce it, the life forms that use it, and the amount that escapes back to the atmosphere. We find concentrations are higher in surface water, lower in deep.

Carbon dioxide (CO_2) on the other hand follows a similar pattern, except that some microscopic animals use it to make their skeletons from one of its compounds. When these organisms die they sink to the bottom. CO_2 concentrations are found to be higher in deep water, lower at the surface level. In spite of this, a balance is maintained.

Calcium, magnesium and sulfur are also important ingredients in this sea soup. Scientists recently discovered something they call a "black smoker" - a chimney-like structure found in several portions of the mid-ocean ridge system. It spews out what looks like black smoke. Instead the dark liquid is a source of calcium. The "smokers" solved yet another puzzle for science: the reason for the marked loss of magnesium and sulfur in the mid-ocean ridge system. The "smokers" remove some of the magnesium and sulfur from the water that circulates through it so that the water coming from the black smoker contains smaller amounts of these elements than that found in surrounding seawater.

Earth's oceans are very complicated chemical systems involving extremely delicate balances that must be maintained. When the rate at which an element enters or leaves the ocean changes, the total amount of that element in the ocean will vary. A steady state can be maintained only when the amount entering equals the amount leaving in a given amount of time. What do you now think takes place in ocean dumping or when vast amounts of spilled oil remains in the sea? Let me quote from a letter written by Jacques Cousteau:

"If we continue to use the sea as our global sewer
and if we continue to disrupt the natural processes of
our biosphere we will undoubtedly bring upon ourselves
catastrophe after catastrophe...."
– Jacques Cousteau

What Cousteau said may be putting it mildly. Think about this. Phytoplankton produce 70% or more of the Earth's atmospheric oxygen through photosynthesis. These plankton gather and use nutrients from the water and could easily be poisoned by the constant, steady decomposition of toxic waste and garbage dumped in the sea. If the plankton are poisoned, it could happen suddenly, according to some theories. It has been known for many years that pollutants slow down photosynthesis in plankton. Is it sensible to take this chance with dumping? Yet we do take this chance. When Rachel Carson wrote so intelligently of the sea, she could not imagine the world's vast oceans, measuring 328 million cubic miles, would be anything but impervious to our actions. Studies